The I Ching & The Genetic Code

The I CHING
&
The
GENETIC CODE

THE HIDDEN KEY TO LIFE

Dr. Martin Schönberger

Foreword by Lama Anagarika Govinda

Translated by D. Q. Stephenson

AURORA PRESS

P.O Box 573   Santa Fe, NM 87504

Copyright 1992 by Aurora Press, Inc..
All Rights Reserved

Originally published in German as
Verborgener Schlüssel zum Leben
Copyright 1973 by Scherz Verlag, Bern and Munich

No part of this book may be reproduced by any mechanical, photographic or electronic process, or in the form of a phonographic recording, nor may it be stored in a retrieval system, transmitted, translated into another language, or otherwise copied for public or private use, excepting brief passages quoted for purposes of review, without the permission of the publisher.

ISBN: 0-943358-37-X

A publication of
Aurora Press
P O Box 573
Santa Fe, NM 87504
USA

The diagrams for the DNA were taken from Jacques Monod's book *Zufall und Notwendigkeit* by courtesy of the Piper Verlag, the I Ching symbol from Richard Wilhelm's book *Buch der Wandlungen* by courtesy of the Eugen Dietrich Verlag. The picture on page two was taken from Ingo Lauf's volume *Das Erbe Tibets* by courtesy of the publishers, Kummerly & Frey

Printed in the United States of America

CONTENTS

Forward to Second Edition 9
Introduction 15
The Opinion of a Mathematician 25
Foreward 27
The Discovery 31
The Genetic Code 37
The Code of the I Ching 45
I Ching as World Formula 59
Binary System and I Ching 63
Method of Transcription 69
Combination of The Genetic Code and the I Ching
in a Single Table 75
Psychological Impediments to an Order of Reality 79
Polarity in the I Ching and the Genetic Code 83
I Ching, The World Code and DNA, the Life Code
A Key? 87
Freedom and Programme in the I Ching 97
I Ching, Law of Chance 101
Chance and Necessity in DNA,
Surrealism and I Ching 105
The Practice of Prediction in the I Ching 111
The I Ching and the I Genetic Code in the
5-Stage Pattern of Meditation 117
Summary of All Reflection 141
Epilogue 147
The Opinion of an Elementary Particle Physicist 153
Bibliography 154
Footnotes 155

In this reproduction of a stone rubbing¹ the couple appears *twice:*

above, the man and woman are turned *towards each other* and the dragon tails are entwined —

below, after union (hieros gamos) they still form an inseparable couple but now the upper parts of their bodies are *turned outward;*

Fu-Hsi holds the set-square in his left hand. As it is used to draw the square, it represents an emblem of the earth, i.e., of the "female principle" (yin), and can hold good as the insignia of the male principle only *after* an exchange of attributes completed during hierogamy². Similarly Nu-Kua, his wife, carries the compasses, the circle-producing emblem of the sky, of the "male principle" (yang).

According to Edouard Chavanne's adaptation of the memoirs of Se-Ma-Ts'ien³, a historical account some 2,000 years old which in many ways reads like mythology but has meanwhile received surprising scientific corroboration, Fu-Hsi and Nu-Kua (Niu-Koa) *"the first married couple"* (and at the same time brother and sister) stand at the dawn of history as it emerges from myth. They have mythical characteristics (dragon tails and wings) and also carry instruments of precision (set-square and compasses). Together these instruments betoken: "order and correct behavior".

Never again in Chinese history do we find such a striking picture of a couple, where stress is laid on the equal excellence of the high cultural achievements of the two partners, which are recorded partly in the form of myths (Nu-Kua's deeds in restoring order to a world out of joint) and are partly of a historical character (Fu-Hsi as the inventor of a system of knot writing, of the eight trigrams and their arrangement, and also of agriculture and hunting; Nu-Kua as the "inventress" of matrimony with an account of her family clan, etc.[3]) — indeed, the couple seem to be a representation of Tao itself in its appearance as **yang** and **yin**.

This book is dedicated to

FU-HSI

and his wife

NU-KUA

the founders of the I Ching and marriage

May this dedication be taken as an epitome, at once lighthearted and serious, of this book and its combination of ancient wisdom with modern science.

Foreword to the 2nd Edition

Since the first edition of this book appeared, a whole area of new territory has been mapped out in the field of molecular genetics; and it is almost impossible to keep track of the many new publications on the subject. Already our ability to manipulate these discoveries has evoked the risk that through negligence, or indeed abuse, deadly viruses such as those of cancer may be incorporated in coli bacilli (our intestinal symbionts) with just as great a danger for the survival of humanity as that caused by nuclear fission. In the spate of discoveries appearing since the fundamental identification of the 64-triplet code one particular one, which was not published until 1975, stands out as being of special importance because it is of equal status with the principle of the DNA code itself. What I am referring to is nothing more nor less than the discovery of the "other half" of the DNA system: the half that is complementary to the material aspect which has so far been the only one defined.

F.A. Popp, instructor at the Radiology Center of Philipps University, Marburg/Lahn, has discovered a complex system, still only partially explored, of vibrations ranging in frequency between ultrasound and ultraviolet light and displaying numerous phenomena such as absorption, reflection, polarization, depolarization, resonance and even a laser function (with a minimum of photons) which is associated in principle with the known chemicophysical structure of DNA, corresponds exactly with its nodes and antinodes, and forms a unity with it. This discovery has vast implications. But that is not all. According to F.A. Popp, the wave character of DNA also implies that between the cells of the body there is a *universal system of communication* operating at far higher speeds than the humoral and neural systems which have alone been known to us hitherto, namely at velocities ranging between those of sound and light. Ultraviolet biosignals "ride" on the spirals of DNA and activate specific codons. Falsification of these signals means cancer; their extinction "puts out the light" of the whole body.

It will be clear that my contention that there is a stream of transcendental, informational power flowing into the DNA is made a great deal more credible and vivid by this discovery than when the first edition appeared. It will also be obvious that reality can be grasped in its totality only if there is a permanent awareness of the living synchronicity of information (codon), behaviour and process in time in the form of a mental equivalent (for which I propose the term "psychon"), and of the two given facts, the DNA substrate and the DNA codon, which should really be called the "somaton". This tripartition corresponds to the mind, soul, and body of Western philosophy. According to Dr. H. de Witt[1] there are close analogies with the three pillars of modern science as shown in the following table:

Biology (DNA)	Western philosophy	Modern science
Codon	Mind	Information
Psychon	Soul	Energy
Somaton	Body	Matter (Mass)

The concepts/qualities/aspects in each line are analogous to each other.

This conception is equivalent to that of the tri-kaya in Tantric Buddhism, the three regions of mind, speech, and body. According to Tantric doctrine these three bodies may be integrated as a unit in a fourth body (sahaya-kaya) of spontaneity and great happiness where the three levels are made one. In the future, science will have to take account of DNA existing as codon-psychon-somaton at one and the same time. This requires, as Jean Gebser shows[2], the replacement of the rational thought dominant for the last 2500 years by a new consciousness in which opposite poles become integrated. This polar thinking and consciousness finds its perfect image, precisely correlated with mental-spiritual-physical reality, in the world system of the I CHING.

The scientist of earlier centuries would certainly have found the complex structure of a television set with its wire connections, coils and capacitors an insoluble riddle without a basic knowledge of electricity and such concepts as plus-minus, electrostatic tensions, the laws of electric current, and induction. And the biogeneticist of today finds the jumble of the DNA code sequences, helixes, and clover-leaf forms an equally insoluble riddle if he confines himself to the field of science and excludes powers, creative formative forces, the polarity of yang-ying, and transcendence. The I CHING, which, by this hypothesis, is coincident with the DNA system, is perhaps the textbook of this cosmic force, static tension and dynamic flux flowing into the matrix of the DNA (as shown in the diagram on page 91).

Dr. F.A. Popp has, I rejoice to say, discussed my hypothesis with his colleagues at his Institute and has written me, that he considers it to be probably correct, because Nature always takes the path that saves most energy. This optimization of energy expenditure is assured by one and the same 64-triplet system at the level of information ("mind") and matter ("body"). *"Since information ("mind") and matter ("body") cross in the genetic code, it may be expected that evolution has selected the most favourable, i.e., the surest and at the same time most economical, principle."*

"The only solution to the problem is to seek understanding of DNA not only in terms of biochemistry and energetics in order to grasp all its fundamental and infinitely wide-ranging significance, but also to tackle the problem, as you have done, with no less energy, from the information theory angle — an approach which is still neglected today. Therefore I can only congratulate you on your pioneering work which is, perforce, still incomplete!" (Quotation from letter).

Nor has proper attention yet been paid to the binary codification and structure of the 64 triplets as revealed in the *periodic order of the amino acids* presented in my table on page 78. It is extremely unlikely, that this computer system embodied in the DNA should as such be left idle and unused. If the hypothesis advanced here is true, then this "keyboard of life", as E. Chargaff, the actual discoverer of DNA, called it, is in fact "played" by vibrations which are exactly matched to all the phenomena of wave mechanics and indeed represent its material counterpart on the analogy of the wave-corpuscular aspect of light.

Both F.A. Popp's discovery of wave systems, which are associated with DNA and also its mathematical structure, have far-reaching consequences for molecular biology. But at the same time they throw an illuminating and explanatory light on the hypothesis of a uniform code system for mind, psyche, and soma. The gulf separating the DNA molecule and an ancient Chinese book of wisdom, the I CHING, once seemingly unbridgeable, now, four years after the first edition, appears to be far less formidable.

May this edition prove a temporary bridge — not an easy one to negotiate — to all those many scientists and laymen who are prepared to integrate the Aristotelian dualism of the rational level of consciousness with the new polar consciousness which my friend and mentor Jean Gebser calls "integral". A bridge between knowledge and wisdom for people who do not suffer "like gypsies on the rim of the universe" (Monod) in the battle of the sexes but seek, united as couples, to form a new society joined together in consciousness of an all-fulfilling polarity.

Introduction

by Lama Anagarika Govinda

There are two reasons which impel me, as a Buddhist and member of a Tibetan order, to accept an invitation to write the introduction to a book which, apparently, belongs to an entirely different field of knowledge, namely modern biology, as seen in the light of an ancient Chinese philosophy handed down to us in the I Ching or "Book of Changes". In the first place, I was moved by admiration for a pioneer work which, free from preconceptions and with the courage of a deep conviction, builds a bridge between East and West, between the early period of human thought and the latest discoveries of science. But the most important reason is because the philosophical principles on which the spiritual attitude of the "Book of Changes" is based, and at the same time the results of the most modern scientific research, are largely identical with the basic ideas of Buddhism and more particularly with the principles of Tantric Buddhism of the Tibetan kind.

The philosophy of the tantras is based on the infinite intrication of all things and living processes whose mutual relatedness and interdependence make the universe into a giant organism. Each part, each individual outward form, contains the whole, is determined by the whole, so that no "thing", no natural process and no living creature exists independently within itself nor can be segregated from the rest but participates in the whole. What we are dealing with here, then, is an organic, i.e., a living unit — not with a mere uniformity or an eternally unchanging substance in contrast to which everything subject to change and alteration is reduced to sheer illusion and unreality and dismissed as valueless. Here unity is not at variance with movement, change, growth and dissolution, evolution and integration which are the characteristics of all living things — like breathing in and breathing out, systole and diastole, and the continual creation and destruction of worlds in the cosmos.

This thoroughly dynamic view of the world, which is opposed to the abstract notion of an unchanging and absolute (and hence unrelated) "being" the concept of "coming into being" (and "going out of being") and could therefore never subscribe to the naive idea of a beginning in time or of a world arbitrarily fashioned out of the void by a creator god, has always been peculiar to Buddhism.

In ancient China, however, there arose a similar view in the notion of "Tao", of *world progress*, of the cyclical movement of the living universe which, like a river, creates law and order through the stability of its orientation and its own innate rhythm, and brings forth all outward appearances and living forms, animates them, and permeates them with meaning. Richard Wilhelm, who has transplanted and interpreted the I Ching and the Tao Te King with genius and, better than any other European before or since, has penetrated to the depths of Chinese thought, therefore identifies the Tao with "MEANING".

It was no surprise, then, in view of these large conceptions of an infinite and dynamic universe, that the meeting of the two great philosophical and cultural streams of Buddhism and Taoism should become one of the most fruitful and humanly appealing events in the whole of man's spiritual history. It was this that impelled me, with a sense of inner necessity, to study the Tao Te King and the I Ching. I was initially acquainted with the latter work through the tradition of Tibetan natural philosophy, astrology, and chronology and was induced to trace back to its origins this ancient system of ideas. This led me to a deeper understanding of the mathematical, geometrical, symbolic and archetypal basic structure of the I Ching which preceded the philosophical and ethical evaluation of this system —

that is, which antedated its use as a book of oracles. To serve the purposes of prediction, first of all the laws of natural events and their application to the human psyche and human action had to be recognized, brought into a uniform system whose sign language and symbols had to be of sufficient general validity, and comprehensibility to operate down the generations, amass additional experience and incorporate it in the system. For it is clear that the "Book of Changes" was not the creation of a single individual, but one of many generations whose experience, accumulated and increasingly crystallized, brought the book to its perfected form. Just as an astronomical prediction is possible only after the constellations have been closely studied and knowledge obtained of their laws of motion based on mathematical calculations, so the basic principles of a universal, that is, a generally valid, world structure had to be created before it became possible to apply them to human life and its situations.

In the process of this application, the purely mathematical establishment of abstract values and clear-cut results, came to be a probability calculus with a steadily increasing range of meanings, which had to be expressed by correspondingly variable and multivalent symbols. But since it is the essence of a true symbol to operate at many levels, that is, to open up a new dimension on the plane of reality, it is clear that these symbols can be correctly interpreted only by someone who knows how to apply them at different levels of reality or consciousness. But since this is impossible for anyone who is inexperienced and unfamiliar with the deeper essence of the symbol, he takes the symbol at its face value. Furthermore, among the existing symbols of the I Ching, various categories must be distinguished which cannot be arbitrarily related one to the other, since they correspond to different levels of reality. In other words, understanding the I Ching does not only depend on philosophical knowledge, but also, and to an equal degree, on a knowledge of symbolic language, which will not yield its secrets to philological inquiry alone. Even such a capable scholar as the English sinologue Legge completely failed to understand the I Ching; he regarded it as nothing but a soothsaying book based on popular superstitions. Unfortunately, in spite of Richard Wilhel's masterly translation, the "Book of Changes" has met with a similar fate today; it has become popular in a way that has subordinated the real meaning of this mighty work to the most trivial ends, and serves to satisfy personal curiosity, or merely fill in an idle hour, instead of fostering world understanding and self-knowledge.

People who have no notion of the inherent orderliness of everything that happens, and of the deep symbolism in which this orderliness is expressed and applied to the human condition, use the "Book of Changes" just like those late Taoists who created a fantastic and mystical folk belief and a popular soothsaying cult out of the profound thoughts of Lao-tse and Chuang-tse. By so doing, they miss the core of the matter, which is the universality of a philosophy to give their life meaning and direction.

Undoubtedly it is possible to recognize the future, or rather, future possibilities, and to control them, but it needs extensive study, an inner suppression of the self, and an intuitive power to which such a discipline gives rise. For intuition is more than just a spontaneous feeling; rather, it is the result of a spiritual growth which derives from long and penetrating preoccupation with a matter or a problem — leading to the identification of all the individual features of the object hitherto remained separate to specific reflection, and making them immediately present to the mind in all their connectedness.

Thus it says in the "Book of Changes":

"Because the wise men of the past thought out the order of the external world, down to its ultimate constituents, and the law of their own inwardness down to its deepest core, they succeeded in understanding fate".

This is a clear exposition of the origin and the way of the I Ching. It is made up of two components: first, objective contemplation and inward assimilation by bringing the mental powers to bear on the processes and laws of the external world, and, second, immersion in the depths of the inward world via the path of meditation, vision, and contemplation by which we not only learn to know ourselves, but also penetrate to the heart of the universe, focalized in the depths of our unconscious mind.

In its recognition of the universality of the human unconscious mind, the philosophy of the I Ching comes into contact with Mahayana Buddhism, whose main doctrine consists in the postulation of the "alayavijnana", of the universal (or "treasury") unconsciousness which is closed to superficial people and therefore unknown to them, but becomes manifest to those who turn inward to themselves. "In it", as the I Ching says, "are the forms and regions of all configurations of heaven and earth."

If we grant this, it is easy to understand that the human consciousness is capable of reflecting the laws of the universe and of reconstructing in its mental processes what has been intuitively grasped, or beheld, in a state of meditative reflection, and that the results of the latest scientific research as in astronomy, nuclear physics or biology were often anticipated by the seers of the distant past and expressed in the symbolic language of their time.

Thus, two and a half thousand years ago, the Buddha spoke of a universe that comprised innumerable world systems and saw clearly that these world systems, of which our constitutes only one tiny individual case, is subject to a perpetual process of creation and destruction, of aeonian evolution followed by aeonian involution, until there is complete integration and re-creation. The naive idea that our earth, or even our solar system, was the center of the universe was first embraced by the anthropocentric thinking of Western peoples and monotheistic religions.

Similarly Buddhism was opposed to the rigid concept of the atom as an indestructible or permanently unchanging substance; and in biology it anticipated Darwin's idea of the evolution of living forms and the gradual evolution of consciousness — from the glimmerings of the animal mind to the highest flights of human thought.

If we bear all this in mind, Dr. Schonberger's astonishing discovery of the identicality of the genetic code and the numerical structure and polarity principle of the I Ching is not only understandable, but also convincing and scientifically acceptable.

And at the same time, the odium attaching to a primitive "oracle", is removed from the function of the I Ching, in foreseeing and shaping the future by the fact that it is a mathematically based program working on the same binary principle as a computer This comparison was drawn some years ago by Jose Arguelles in a very informative article (in "Main Currents of Modern Thought", January 1969) in which he wrote:

"The I Ching functions like a computer, and its function depends on its being programmed in conformity with truth. The truth (or correctness in terms of the facts) of the programming depends on how the person consulting the book reacts to the utterances of the "Book of Changes".

In other words, the I Ching functions only to the extent that the consultor accepts the rules and laws laid down in the book, and applies them to his own situation." The I Ching is a kind of "psychic computer", i.e., a combination of subjective and objective factors, psychic and mathematical data, in which the correct appraisal of the former is essential for the functioning of the latter.

Using his own line of approach, the author of this book has obtained the same result independently by likewise replacing the much misunderstood word "oracle" by *programming*. Such a programming of the living process is provided by Nature in the form of the *genetic code,* and may thus also be conceived as a programming of the fate of each living creature. As the author himself says in this connection,

"From the outset the Chinese has had no doubt that the creation of the world from the primary poles must have been decreed by natural necessity or "fate", although with variable courses and developments, and indeed he could never have conceived otherwise."

As a book of fate of this kind, the "Book of Changes" has for thousands of years acted as a catalyst to human thought an a guide to human action. The reason why this was possible must be sought in the general validity of the underlying observation and formulations in which the dynamics of all living things and their inherent conformity with natural laws are equally reflected. It is this conformity with natural laws that gives stability to change. Instead of seeing only death and annihilation in change and, out of this negative attitude, evolving its opposite, namely, the eternally unchanging and immutable, and raising it to an ideal, man in the earliest Chinese civilization had the courage and insight to affirm this eternal change and to win through to the knowledge that change is not the contrary of stability, but is indissolubly bound up with it. In other words the early sages of China did not fall victims to dualistic thinking — which makes change the inexorable enemy or opponent of stability — but recognized the polarity of these two sides of reality in the sense that, although they are opposed, they necessarily involve each other.

Dualism* and polarity are two concepts which are worlds apart yet, unfortunately, they are very often confused with each other — mainly by those who make the idea of an absolute unity conceived as the sole reality into an exclusive ideal in contrast to which multiplicity, diversity, differentiation and individualization appear as a "lapse" from absolute reality and are disparaged as mere illusion. The difference between dualism and polarity is that dualism sees only the irreconcilable opposites and leads to prejudiced evaluations and decisions which sunder the world into equally irreducible opposites; whereas polarity is born of unity and includes the concept of wholeness: the poles are complementary to each other, are indissolubly involved in each other, like the positive and negative poles of a magnet, which imply each other and cannot be separated. The flaw in dualism is that we want to accept only the one side of things or life processes, namely, the one that is agreeable to our wishes or our ideals, and even more, to our illusory "ego" and everything with which it is identified.

Thus the concept of immutability is confused with that of duration, which might perhaps be better expressed as con-

---

*Taoism (I Ching) and Buddhism are also at one in declining dualism. The Buddha himself declared: "The World, O Kaccana, is devoted to dualism, to the "It is" and the "It is not". But, O Kaccana, for him who in accordance with reality and in the fullness of wisdom, recognizes how things arise in the world, there is no "It is not" in this world. And for him, O Kaccana, who in conformity with reality recognizes in wisdom how the things of this world pass away, there is no "It is" in this world" (Samyutta Nikaya II, 17).

tinuity. In the definition of the I Ching:

"Duration is a condition whose movement is not exhausted by obstacles. It is not a state of rest (in the sense of absence of motion), for mere standstill is retrogression. Hence duration is self-renewing movement of an organized, integrated whole which proceeds in harmony with immutable laws."

In this sentence the central idea of the ''Book of Changes is reduced to the shortest formula, one which will also receive the assent of modern science, as the author of this book has convincingly shown with an abundance of significant details and parallels. May the ideas advanced in this book, representing as they do both the most ancient and the most modern results of human research, be of service to many in stimulating and deepening their own philosophy.

Kasar Devi Ashram,
Kumaon Himalaya, India,
31 January 1973

The Opinion of a Mathematician

Every reader, whether in the scientific or philosophical camp, will be surprised to find the opinion of a mathematician at this place. For it is our habit, inculcated by a strict education, to assign our knowledge to two mutually exclusive categories: to the category of the exact sciences, and to that of philosophy. And we are also accustomed to accepting the claim of science as absolute: It alone is, in principle, capable of describing all the knowable phenomena of this world and of explaining them. It is idle to argue of the correctness of this statement. Invariably the scientist is forced to admit that there is still a vast residue of the inexplicable, and time and again the philosopher must witness how the scientist arrogates one field after another to himself. And so it is mainly a matter of our education, and the system into which we were born, whether we affirm or deny the correctness of what is written in these pages.

What is it that makes this book and the system of the I Ching so fascinating? Firstly, it is the abstract formalism on which the I Ching is based, and in this respect the I Ching is not dissimilar to a mathematical theory. And secondly the fact that our power of reasoning in the sense of logical thought, i.e., our ability to acquire scientific knowledge, is contained in the I Ching system as one of the eight mental faculties; in other words, our scientific mode of thinking also has equal title to a place in the I Ching system. Hence there is no need for us to dissociate ourselves from any part of our nature. However, absent from the I Ching, but forming the whole basis of our thinking, is the concept of duality, the principle of dichotomizing into opposite and mutually exclusive categories. Its place is taken in the I Ching by the principle of polarity, i.e., the principle of uniting two apparently contrary categories. Who would blame the author for pointing out the enormous difficulties which have accrued to physics precisely because of this dualistic thinking? It must therefore be a source of great satisfaction for a devotee of the I Ching when a physicist require elementary particles to be regarded as both corpuscles and waves — a demand which is in fact an open invitation to polar thinking!

For this reason, it seems to me, it is eminently worthwhile for the scientist to read and inwardly digest these pages. What is so profoundly thought-provoking is undoubtedly the astonishing similarities between the genetic code and the I Ching code, which are described at the end of the book. And then one is surprised to find that the I Ching contains a philosophy of life which is erected on a very formal basis, and yet at the same time, does not lend itself in the least to transformation into an ideology and hence to misunderstanding. These two facts alone must make it easy for the scientist to approach the author's thinking with an open mind.

One word of warning before concluding. This book cannot undertake to provide a synthesis between the thinking of East and West. It necessarily contains many points which will provoke the scientific thinker. For one reader the striking affinity between the I Ching symbol and the DNA double helix may seem due simply to the interplay of random factors; another will want to banish the oracle technique of the 64 I Ching hexagrams to the realm of magic. Indeed every reader may find points which he regards with scepticism. Nevertheless, leaving all this aside, the book does contain ideas which are very well worth being pursued further, particularly by scientists.

Rudiger Hauff
IBM Stuttgart

Foreword

There are milestones in the history of mankind which loom enormously large and always betoken a new phase of human development. We can form only some dim conception of the number of such fateful events in prehistoric times nad attempt to reconstruct some of them in our imaginations. There was, for instance, the discovery of fire which perhaps marked most clearly the transition from the prehuman to the human field. The invention of the wheel, the lever, agriculture (which shows a curious affinity with the ability recently acquired to control human fertility), printing and many others. In spite of the vast number of inventions and additions to our knowledge, our more remote descendants might pick out two such milestones as being the greatest discoveries of our day and as having the greatest consequences for posterity: the fissionability of the atom and, in the biological field, the unitary origination, organization, preservation and propagation of plant, animal and human life by means of the genetic code, even though the latter has still not made such a dramatic impression on the human mind.

It was probably much easier for the layman to grasp the importance of nuclear fission. Unless he has learned about it as a reader of scientific literature, the educated layman is likely to have heard only vague rumors of the discovery of the genetic code, its carrier substance, and all the processes which it involves. For the biologist, of course, the great importance of this knowledge is clear, and no doubt research work has been going ahead at top speed in various biological centers for protein research all over the world, in order to harvest the fruit of this seed as early as possible. Indeed, the journals even tell us that, after the Promethean deed of discovering the genetic code had been performed, it is now mainly a matter of tackling the bewildering variety of special problems that have cropped up. The conferment of the Nobel prize on Watson and Crick in 1952 seemed to be an adequate tribute to the importance of the discovery, and it is not customary to go on being astonished about discoveries which have already become an established part of scientific knowledge.

This small work by a physician, who is no more a scientist and philosopher than many of his colleagues, and who may even be guilty of errors of style and form, was born of a desire *not* to rush forwards too precipitously. For it was precisely the doctor with an interest in natural philosophy — the kind of physician honored by Hippocrates — and also other friends of natural philosophy — who were more profoundly stirred by the first fragmentary press announcements of the basic facts about the genetic code, and everything pertaining to it, than the pure specialists with no connection to natural philosophy. For it flashed upon him like a thunderbolt that, beyond any shadow of doubt, this curious programming of all life processes by means of 64 code words, each consisting of *three* of *four* "letters" — a system, that is, which struck most of those hearing of it as rather peculiar and odd — was something already familiar to him down to the smallest details. It reminded him, in many of its details, of the complicated structure of an admittedly relatively little known philosophy which he first came to know with similar feelings of puzzlement and astonishment, because of its highly individual character and strangeness. This book, the I Ching, perhaps the oldest book in the world, also has 64 signs consisting of *four* "letters", of which only three are used at a time, at least in the version dealt with here.

The "basis" of the genetic code is formed by the *plus* and *minus* threads of DNA. The I Ching rests upon the two basic principles of yang and yin, which we need not hesitate to term *plus* and *minus* poles. At the same time, it is a philosophical system which, when its sometimes abstruse forms and formulae are seen in the cold light of modern knowledge, seems to scintillate as if it had myriad facets. It is claimed to be a law of universal validity which is not only law, but also said to be the origin of the whole visible world, and to be operative in the finest details of its patterns and its turnings of fate — while at the same time constituting a code of conduct for the best possible form of interhuman relationships. Anyone knowing the work may have laid it aside with a shake of the head, and yet time and again, retrieved it from the recesses of the library in order to consult it. Whatever else, it is one of the most curious books in *any* library, however extensive; it is like an erratic block, a prehistoric find, which refuses to fit into a well-ordered landscape.

The astonishment and bewilderment evoked in the author by these very curious parallels between two such mutually remote systems has never since left him, and so, in 1969, when a medical magazine happened to print an article on the I Ching for the general reader, he felt called upon to publish an account of the comparability of the two codes. The reprint appearing at that time is reproduced here and followed by more detailed notes on the genetic code and the I Ching for the sake of greater clarity. The two codes are shown in a provisional table written into each other. Finally, various attempts are made to fit this strange and surprising find into the existing scheme of things.

It was only when this book had been completed that the author came across the essay by Dr. Marie-Louise von Franz "Symbol des Unus Mundus"[1] in which there is a first intimation of the relations between the I Ching and the DNA code. Being still uncertain in his judgment of these curious parallels, the author submitted his work to a man who, in his writings, has time and again referred to the need and the possibility of building a bridge between the Western and Eastern mind, and from whom he expected frank criticism and appraisal. The reply was confirmatory. This man regarded the "discovery" as "extraordinary and unique", but made it clear that the first publication entirely failed to cover the subject adequately. A very cordial word of thanks is due to Jean Gebser here. The foundation for this gratifying collaboration was, of course, laid a long time ago in the efforts made by the author over many years to find a non-dualistic intellectual attitude based on polarity. What he sought could be found in the I Ching in a noble and unique efflorescence of the human spirit. The present report on the coincidence and minifestation of polar structure in the basis of all life (DNA), may well portend a comprehensive realization of polar thinking and experience in the future.

While this book was still being prepared Monod's book "Chance and Necessity"[2] was published. It contains many ideas about the genetic code which are formulated in strictly dualistic terms. The author therefore felt it incumbent upon him to reply from the viewpoint of the I Ching, more especially in view of what Monod calls the "ultimately unsolved riddle of the origin of the genetic code"[3] and of the "Book of Changes" as a *reply* to the riddle and its *solution*.

This book is therefore an attempt to call attention to a remarkable discovery, and to encourage qualified experts to examine it more closely.

The Discovery[1]

I Ching — The Book of Changes

and

The Genetic Code — The Book of Life

E. H. Grafe's communication: "I Ching", in No. 5/69 of the Zeitschrift fur Allgemeinmedizin — Der Landarzt, must no doubt have contained information which greatly surprised the majority of readers.

What is an ancient Chinese book of oracles doing in a medical journal?

It was probably impossible, within such a narrow compass, to show that the I Ching does represent in essence an extract of Chinese natural science. This aspect of the I Ching is presented with much greater clarity in Grafe's books, to which the reader's attention is expressly drawn. Perhaps this first impression will be corrected by setting up, on the printed page, the astonishing parallels between the natural science of the I Ching and the latest discoveries of nuclear genetics.

Natural Science "FORM" The Genetic Code: "The Book of Life" *John Kendrew*	Philosophical Theory "CONFIGURATION" "The Book of Changes" Compendium of natural philosophy from ancient China, compiled by Fu-Hsi and edited by Confucius
1. Discovered 10 years ago, has existed since life began. All	1. All processes of living development throughout na-

the vital processes of all living creatures whose structure, form and heredity are programmed in precise detail *universal claim*

2. The basis is the plus and minus double helix of DNA

3. Four letters are available for labelling this double helix: A-T, C-G (adenine, thymine, cytosine, guanine), which are joined in pairs

4. Three of these letters at a time form a code word for protein synthesis

5. The "direction" in which the code words are read is strictly determined ($\rightarrow$)

6. There are 64 of these triplets known whose property and informative "power" has been explored. One or more triplets program the structure of one of the possible 22 amino acids; quite specific sequences of such triplets program the form and structure of all living creatures, from the amoeba to the iridescent peacock's feather

ture are subject to *one* strictly detailed program (*universal* physical, metaphysical, psychological, moral *claim*)

2. The basis is the manifestation of the world principle in the primal poles yang (—) and yin (- -)

3. Four letters suffice for life in all its fullness.
   7 = resting yang
   9 = moving yang
   8 = resting yin
   6 = moving yin

4. Three of these letters at a time form a trigram, a primary image of the 8 possible dynamic effects.

5. The "direction" in which the trigram is read is strictly determined ($\uparrow$)

6. There are 64 double trigrams precisely designated and described by Fu-Hsi (3000 BC) in very vivid and precise images of highly specific dynamic states (e.g. "breakthrough", or "oppression") with in each case 6 possible variations of this state and subsequent transformation into another one of the 64 hexagrams — a program of fate, as it were, in which each individual is at all times placed to operate the "switch" of fate from which point onwards the "train"

7. Two of these triplets have names: "beginning" and "end". They mark the beginning and end of a code sentence of some length.

7. Two hexagrams have names: *before* completion — *after* completion (frequently opening and closing "melodies of fate" in the oracle).

Table of the
Genetic Code
(Watson Crick)

First		Second			Third
	U	C	A	G	
U	Phe	Ser	Tyr	Cys	U
	Phe	Ser	Tyr	Cys	C
	Leu	Ser	ochre	?	A
	Leu	Ser	amber	Tyr	G
C	Leu	Pro	His	Arg	U
	Leu	Pro	His	Arg	C
	Leu	Pro	Gln	Arg	A
	Leu	Pro	Gln	Arg	G
A	Ile	Thr	Asn	Ser	U
	Ile	Thr	Asn	Ser	C
	Ile	Thr	Lys	Arg	A
	Met	Thr	Lys	Arg	G
G	Val	Ala	Asp	Gly	U
	Val	Ala	Asp	Gly	C
	Val	Ala	Glu	Gly	A
	Val	Ala	Glu	Gly	G

Fu-Hsi's Table of the I Ching

Upper Trigram / Lower ▼	☷	☶	☵	☴	☳	☲	☱	☰
☷	2 Kun	23 Bo	8 Bi	20 Guan	16 Yü	35 Dsin	45 Tsui	12 Pi
☶	15 Kien	52 Gen	39 Gien	53 Dsien	62 Siau Go	56 Lü	31 Hien	33 Dun
☵	7 Schi	4 Meng	29 Kan	59 Huan	40 Hiě	64 We dsi	47 Kun	6 Sung
☴	46 Schong	18 Gu	48 Dsing	57 Sun	32 Hong	50 Ding	28 Da Go	44 Gou
☳	24 Fu	27 I	3 Dschun	42 I	51 Dschen	21 Schi Ho	17 Sui	25 Wu Wang
☲	36 Ming I	22 Bi	63 Gi dsi	37 Gia Jen	55 Fong	30 Li	49 Go	13 Tung Jen
☱	19 Lin	41 Sun	60 Dsiě	61 Dschung Fu	54 Gui Me	38 Kui	58 Dui	10 Lü
☰	11 Tai	26 Da Tschu	5 Sü	9 Siau Tschu	34 Da Dschuang	14 Da Yu	43 Guai	1 Kien

True, further comparison leads to more parallels but — in view of the extreme disparity of the "languages" (Fu-Hsi's and Watson and Crick's) — also to many specific problems.

Thoughts evoked by these striking parallels:

Is this similarity fortuitous? It seems hardly likely.

The fact that the precise programming of the lifelong identity of all living creatures and their heredity is determined by the genetic code in 64 words and 3 letters (out of 4 possible ones) and has been, *without our knowledge*, since life began, is one of the greatest and most momentous of discoveries. It makes good sense and sheds light on whole worlds of interconnected relationships.

The knowledge that all other developments and patterns of fate and fortune are subject to the same strict law of cause and effect and programming in a system comprising 64 potential states, with 6 possible emphases, and a mathematically infinite multiplicity of transformations into any one of 64 other possible states (etc.) (even where nothing is known of the I Ching law) comes as an extraordinary shock to the European mind. It will, no doubt, because of its very strangeness meet with sharp criticism, suppression, denial and belittlement "because there cannot be what there should not be" — after 2,500 years of Western philosophy, or more precisely, since Aristotle.

And yet we shall not be able to evade the question: Are both "books" manifestations of a common principle? Is what is involved here perhaps one universal code which was discovered 5,000 years ago by the Chinese — and 10 years ago by Watson and Crick?

In other words: Is there only one spirit whose manifestation (= information?) must *of necessity* find its expression in the 64 words of the genetic code on the one hand or in the 64 possible states and developments of the I Ching (including, be it noted, all aberrations and maldevelopments) on the other?

One law running through the whole of nature in all its diverse physical, spiritual, intellectual, and moral processes as determined by fate?

Here, I feel, something is glimpsed as through the eyes of the physician: we are suddenly vouchsafed a scientific insight into a world, whole and sound, unriven by schizoid discord, where physics and metaphysics are one — as for the pre-Socratic philosophers (enantiodromia) — or in the concluding words of Grafe's book: into "security, calm, happiness".

The Genetic Code

Although the basics of heredity have been known since the chromosomes were discovered, the discovery of DNA (deoxyribonucleic acid) as the carrier of genetic information by Watson and Crick in 1953 marked a revolutionary advance. These scientists were awarded the Nobel prize in 1962 in recognition of their work. A very simplified version of the Watson-Crick model will suffice for our purposes. Readers are referred to the technical literature, particularly James D. Watson's own exciting account "The Double Helix", for more information. DNA, a chain-like molecule of great length and high molecular weight, is a double strand twisted like a spiral staircase, a "double helix" with plus and minus strands, which represents the matrix of the genetic message. The double strand itself consists of two chains of alternating units of phosphoric acid residues and deoxyribose (a simple sugar) combined in a simple buildingblock system. Both strands are joined at regular intervals as if by the rungs of a rope ladder, each rung consisting of a pair of bases. There are four bases: thymine (T) (replaced by uracil (U) in the "transcript") always paired with adenine on the opposite rope of the ladder, and adenine (A) always paired with the opposite thymine (T) which is complementary to it. Similarly cytosine (C) is always paired with guanine (G) and guanine (G) with cytosine (C).

Thus A, G, C, and T are the "letters" of the code, paired with T, C, G and A of the parallel strand of the double helix, and, because of their chemical and spatial structure, they all fit together precisely like the elements of a zipper. Laborious investigations have revealed the surprising fact that three of these letters, i.e., a sequence of bases, always signify a code word in an endless sequence of "words" (with punctuation) on the substrate of the double strand. By "signify" is meant that the code

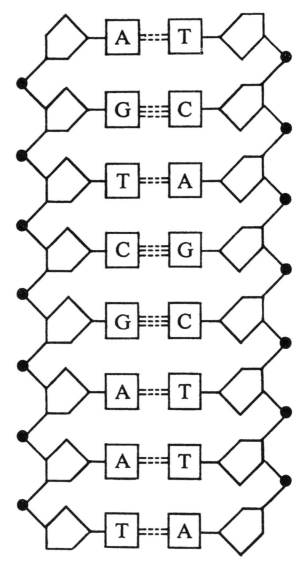

Diagram of the Genetic Code

word is an instruction for putting together (synthesizing) a compound and is not the product itself. According to the laws of mathematics, 64 such combinations, known as triplets, are possible and have in fact been found and recognized, and their meaning, which is fixed by natural laws, has been read. One or more of these words: A-A-A, A-C-G, G-C-A, etc., etc., represents the information and instructions needed for the synthesis of an amino acid, one of the protein building-blocks of the body. Thus one very precisely defined sequence of hundreds of such triplet sequences is specific for the protein structure of one very precisely defined part of a living creature. It is only because of these precisely formulated instructions, which must always remain invariable, that the same protein product is always synthesized. The sum total of all these code words is therefore synonymous with the "blueprint" for producing a whole specific plant or animal body with all its characteristics. In uniovular twins, the way characteristics are reproduced with mirror-image accuracy is striking in the extreme. In a virus, a few hundred code words are sufficient, whereas man needs billions. If all the 48 chromosome-DNA-units were lined up, each would be equivalent to a closely-printed lexicon of 20,000 pages. Assuming the double strand, which is twisted billions of times, could be untwisted, it would be about 1.3 meters long. And this strand with all the necessary building instructions is present in every undifferentiated cell nucleus.

In addition to this function of the living cell by which the blueprint is programmed, and the genetic message maintained unchanged throughout life, (old age is now understood to be due in part to gaps or damage in the blueprint) there is a second function, namely, the reduplication and inheritance of this blueprint. To this end use is made of the structure of the two strands of the "rope ladder", which are complementary to each other in a remarkably precise manner. They can open like the

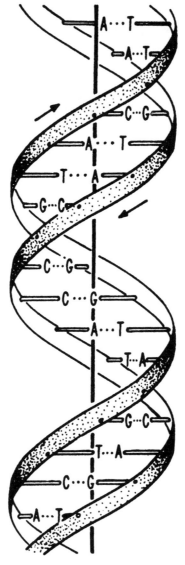

Double Helix of DNA

two halves of a zipper, and each half of the ladder synthesizes a new complementary half of DNA without (under normal circumstances) the information being falsified in any way. Nevertheless, such errors as occur, are particularly serious because they continue to be reproduced ad infinitum in subsequent reduplications (hereditary diseases, cancer).

There is one more detail we need in order to understand the table on p. 44, where U (uracil) is to be read for T (thymine). To translate the DNA code strand into the language of protein synthesis, use is made of a single-strand ribonucleic acid which resembles one half of a zipper and is known as messenger RNA. In RNA, the letter T (thymine) is replaced by U (uracil). The code was discovered in this mediator between DNA and protein — this matrix for ordering the amino acids forming the building blocks of polypeptides — and therefore in the subsequent table, as in the case of RNA, T (thymine) is replaced by U (uracil).

There are 20 amino acids; some of them may have molecular variations (thus making a possible total of 24?), in most cases several code words are needed to synthesize one amino acid. Three code words serve for punctuation: "amber" UAG and "ochre" (UAA) and AUG for the beginning and end of a genetic "sentence."

There are already numerous indications that information of a memory-like character, other than that imprinted on the DNA, is conceivable and that, just like the blueprint, behaviour patterns and instinctual mechanisms are inherited because they are "written down" in the DNA for transmission. This area of transition into what are normally non-material fields, is of particular interest for our study, for it might be said: Fate, length of

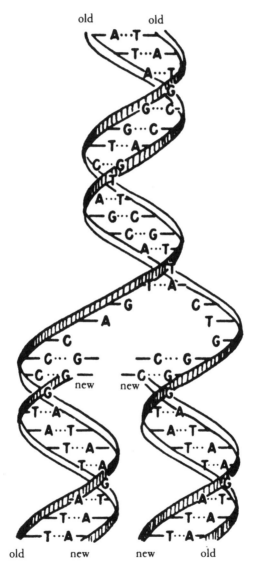

Diagram showing replication of DNA

life, sexual behaviour, the nest-building rites of birds and the subtlest variations of their brood care, indeed all the vital manifestations of every creature and (how far? and how far not?) of man himself are determined, in advance, in every cell of the body by means of the strictly constant DNA codification, which is programmed and laid down as "necessary". Does this non-material element extend far beyond our materialistic perspectives? Is it not possible that in the DNA code — a *script* with significance, meaning, expression, information, impulse and life — we can see and recognize firsthand both spiritual and material structures erupting into the *hyle*, the material? In all events, since the discovery of the genetic code and its 64 words, the fact can no longer be ignored that non-material information is inherent in the nucleus of every living creature as if it were a computer, and that it shapes, maintains and increases all forms of life with unimaginable dynamic force. For our dead DNA model does not show that DNA is continually being replicated, that it is *vitally alive*. If there are, say, 300,000 turns in the helix, then about 15,000 windings a minute must be untwisted for duplication to be complete in 20 minutes (the normal time) — an almost inconceivable process which nevertheless takes place in the cells of every living creature every time a cell divides.

As knowledge continues to snowball, these minimum basic facts had to be explained, however briefly and fragmentarily, since they are largely terra incognita for many educated readers, even though they now figure in the syllabus of secondary schools. **One** system of universal application, constituting in fact the contents, laws and program of every form of life, of the whole animal, vegetable and human integument of the planet in its myriad variations, can be seen here for the first time in the full sweep of its significance — the "book of life".

		Second Letter				
		U	C	A	G	
First Letter	U	UUU ⎤ Phe UUC ⎦ UUA ⎤ UUG ⎦ Leu	UCU ⎤ UCC UCA Ser UCG ⎦	UAU ⎤ Tyr UAC ⎦ UAA OCHRE (STOP) UAG AMBER (STOP)	UGU ⎤ Cys UGC ⎦ UGA Start UGG Tryp	U C A G
	C	CUU ⎤ CUC CUA Leu CUG ⎦	CCU ⎤ CCC CCA Pro CCG ⎦	CAU ⎤ His CAC ⎦ CAA ⎤ GluN CAG ⎦	CGU ⎤ CGC CGA Arg CGG ⎦	U C A G
	A	AUU ⎤ AUC Ileu AUA ⎦ AUG Met = Start	ACU ⎤ ACC ACA Thr ACG ⎦	AAU ⎤ AspN AAC ⎦ AAA ⎤ Lys AAC ⎦	AGU ⎤ Ser AGC ⎦ AGA ⎤ Arg AGG ⎦	U C A G
	G	GUU ⎤ GUC GUA Val GUG ⎦	GCU ⎤ GCC GCA Ala GCG ⎦	GAU ⎤ Asp GAC ⎦ GAA ⎤ Glu GAG ⎦	GGU ⎤ GGC GGA Gly GGG ⎦	U C A G

Table of the Genetic Code

List of the amino acids and their abbreviations in the above code

Ala = Alanine
Arg = Arginine
Asp = Aspartic acid
AspN = Asparagine
Cys = Cystine
Glu = Glutamic acid
GluN = Glutamine

Gly = Glycine
His = Histidine
Ileu = Isoleucine
Leu = Leucine
Lys = Lysine
Met = Methionine
Phe = Phenylalanine

Pro = Proline
Ser = Serine
Thr = Threonine
Trp = Tryptophan
Tyr = Tyrosine
Val = Valine

THE CODE OF THE I CHING

Difficult though it may be to convey a basic notion of the DNA-RNA code, it is even more difficult to describe the I Ching, perhaps the most ancient book in world literature. In all events, C. G. Jung's commentary in "The Secret of the Golden Flower"[1] brought it to the notice of psychologists. During many years of painstaking work with eminent Chinese scholars, R. Wilhelm translated the book I Ching "The Book of Changes" into German[2], and it was only when the text had been retranslated into Chinese, and yielded the same sense, that the German version did, was it held to be reliable. This is mentioned in order to exclude from our consideration various German editions of the I Ching which, while based on R. Wilhelm's translations, dispense with his painstaking accuracy and contain unwarranted simplifications and changes of meaning which mar their quality.

There is no doubt that it would be much easier to present the meaning of the I Ching to European readers and students free from the patina of nearly thirty centuries. Yet it would also be a falsification. We shall therefore simply give the readers some information about the I Ching by illuminating its various aspects as — to use a modern term — a "world formula" with a cosmogony, a generally valid theory of the origination of the visible world from a primary principle, the Tao, which transcends the power of words to describe. The manifestation of this principle, its emergence, is always polar. The world poles are called yang and yin; they condition and supplement each other, unite, appear in a state of rest and motion just as, say, electricity (with a plus and minus pole) can be static or flowing. However, the world of appearances is not conceived as a vague mixture of polar quantities but as being organized on mathematical laws, on polar quanta which spring from the potentially creative "zero" (in reality the Tao).

In the I Ching the "positive" pole is represented by a straight unbroken line which, placed on the previous blank of the paper, posits an above and below, a front and a back, a right and left:

Oneness: ▬▬▬▬▬▬▬▬▬▬   "The ridgepole"[3]

In the polar structure, one "concept" is subsumed by the whole principle: oneness, yang: positive — male — heaven — active — aspiring — sun — south — bright — firm.

In the later Tao-Te-King by Laotse, we read in the translation by Prof. Siegbert Hummel[4]

"Tao posits (at the same time as itself) oneness, oneness posits (at the same time as itself) twoness. (Yang and yin) posit (at the same time as themselves) threeness. Threeness posits (in and at the same time as itself) all creatures."

Now *twoness:* ▬▬▬▬ ▬▬▬▬

Yin signifies: negative — female — earth — passive — sinking — darkness (more specifically: the shadowy, i.e., conditional upon light!) — north — soft. Let it be said straightway (and not thoughtlessly "turned round" in our manner as seen in some translations):

In the I Ching the south is at the "top", the north at the "bottom". As will be seen below, it is essential to preserve this cosmic axis! East and west are also "changed round", compared with our normal practice. There is a second misunderstanding: the poles must not be conceived as being dualistically invariable, but rather as a continuous transformation and transition of these forces, the transformation being partly a continual change from one to the other, partly a closed cycle of unified complexes of events such as day and night, summer and winter.

This changing polarity is clearly depicted in the T'ai-chi. Here the two complementary poles are inscribed in a circle, each fitting snugly into the other and containing the (mutable) seed of the counterpole.

"T'ai-chi"

Transcribed into the line symbolism, the long unbroken line ─────────── represents the light pole which also symbolizes the male, the creative and heaven, whereas the broken line ───── ───── expresses the dark pole, which symbolizes the female, the receptive and the earth. "However, the need for greater differentiation seems to have been felt at an early date and the simple lines were combined in pairs." (R. Wilhelm, "The Book of Changes").

These are the four "letters" of the I Ching code.

The following detailed quotation is taken verbatim from R. Wilhelm "The Book of Changes":

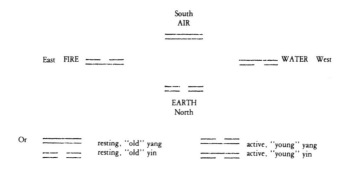

> *"To each of these combinations a third line was then added. In this way, the eight trigrams came into being. These eight trigrams were conceived as images of all that happens in heaven and on earth. At the same time, they were held to be in a state of continual transition, one changing into another, just as transition from one phenomenon to another is continually taking place in the physical world. Here we have the fundamental concept of the Book of Changes. The eight trigrams are symbols standing for changing transitional states; they are images which are constantly undergoing change. Attention centers not on things in their state of being — as is chiefly the case in the Occident — but on their movements in change. The eight trigrams therefore are not representations of things as such but of their tendencies in movement. These eight trigrams came to have manifold meanings. They represented certain processes in nature corresponding with their inherent character. Further, they represented a family consisting of father, mother, three sons, and three daughters, not in the mythological sense in which the Greek gods people Olympus, but in what might be called an abstract sense; that is, they represented not objective entities, but functions."*

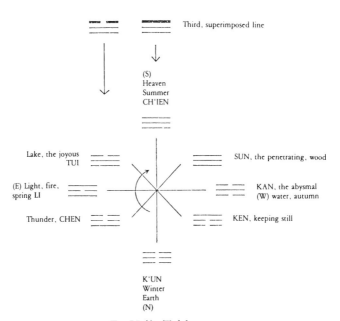

Fu Hsi's Table

If we take these eight symbols on which the Book of Changes is based one by one, we obtain the following order*:

	Name	Attribute	Image	Family
☰	CH'IEN, the creative	strong	heaven	father
☷	K'UN, the receptive	devoted	earth	mother
☳	CHEN, the arousing	movement	thunder	first son
☵	K'AN, the abysmal	dangerous	water	second son
☶	KEN, keeping still	resting	mountain	third son
☴	SUN, the gentle	penetrating	wood, wind	first daughter
☲	LI, the clinging	light-giving	fire	second daughter
☱	TUI, the joyous	joyful	lake	third daughter

The sons represent the moving element in its various stages; beginning of movement, danger in movement, rest, and completion of movement. The daughters represent the element of devotion in its various stages: gentle penetration, clarity and adaptability, and joyous tranquillity. In order to achieve a still greater multiplicity, these eight images were combined with one another at a very early date, thus giving a total of 64 signs. Each of these 64 signs consists of six lines, either positive or negative. Each line is thought of as capable of change, and whenever a line changes, there is a change also of the situation represented by the given hexagram. Let us take for example the hexagram *K'UN*, the Receptive, earth: ☷ ☷

It represents the nature of the earth, strong in devotion; among the seasons, it stands for the late fall, when all the forces of life are at rest. If the lowest line changes, we have the hexagram *FU*, Return: ☷ ☳

It represents thunder, the movement that stirs anew within the earth at the time of the solstice, "the return of the light".

In this Primal Arrangement of Fu Hsi, following the earliest observations of nature, *four pairs* are always shown in this sequence which, as a pattern, must be conceived in this way and no other; intermingling, i.e., entering each other, polar, in reciprocal contact and interrelatedness. This is the mnemonic verse attributed to Fu-Hsi[5].

"Heaven and earth determine the direction. The forces of mountain and lake are united. Thunder and wind arouse each other. Water and fire do not combat each other. Thus are the eight trigrams intermingled. Counting that which is going into the past depends on the forward movement. Knowing that which is to come depends on the backward movement. That is why the Book of Changes has backward moving numbers."

---

* All I Ching trigrams and hexagrams must be read from the bottom to the top.

The natural direction of rotation is clockwise and shows the sequence of the year. However, it seems very strange to us to say that, knowing that which is to come, depends on backward movement. We have a very long way to search in Western science before we find the theory of "backward" movement, time reversal, the disappearance and appearance of plus- and minus-charged particles, and the calculation of future states, until there is mention of anticlockwise rotation. In fact, we find it only when we come to the results and theories of atomic physics — and of the DNA double helix. During the intervening thousands of years of natural science, there is no mention of plus and minus, time reversal, retrograde movement, clockwise and anticlockwise movement, conversion of energy into matter — except in the natural philosophy of the I Ching (no gods are needed!) with its natural forces, form-giving immaterial principles, ideas, ordering tendencies, cardinal points, and stereonomies. Here, in this ancient and archaic garb, we find an early version of an exact natural science! And one point more. If modern physics with its formulae is incapable of ordering mental phenomena appropriately, and in conformity with natural laws, this possibility does exist in the psychological aspect of the I Ching.

To recapitulate, here are the aspects of the I Ching we have discussed:

1. cosmogony (Tao - yang - yin),
2. the moulding of all existence (spirit - mind - matter) by the eight primary images and their 64 manifestations in the spatiotemporal aspect,
3. the aspect of right-hand rotation (moving in a clockwise direction like the signs of the seasons in the primary trigrams) and the aspect of backward (i.e., anticlockwise or leftward movement) rotation which strikes us as a very disconcerting and, as it were, proscribed form of movement in our experience.

According to the I Ching doctrine of spatiotemporality, there should therefore also be, unfolding from the seeds of the eight primary trigrams (the "world" reading rightwards), a backward path, contrary to the natural order of events, through which the seeds can be recognized, the past understood and the law-governed development of the future predicted — a path which is open to the wise man through his intuitive insight into the course of nature, in accordance with the primary trigrams and their 64 combinations, in six possible steps at a time.

This must suffice for the moment as an introduction to the use of the I Ching as a book of wisdom — with a very curious methodology based on what appear to be fortuitous manipulations.

But more of this later.

Let us now return to the psychological and biological aspect of the primary images. Together with deeply versed students of the I Ching, Lama Anagarika Govinda and Jean Gebser, I tried to find superordinate terms for the four paired "intermingled" primary trigrams of the I Ching by which to refer to their psychological and organic aspect. The terms should be in harmony with the R. Wilhelm translation, should avoid falsifications of a dualistic and exclusive nature by, say, the introduction of current and commonly understood concepts of European psychology such as Will, Reason, Feeling, etc., and yet still suggest the unity of word and image found in the Chinese text. They should therefore "accord" and represent the reality of the four interlinked (this way and no other) polar pairs which are at one and the same time spiritual, mental and physical in appearance and function.

Under no circumstances should the validity of the eight "powers" be excluded, even in the material field. But here our vocabulary, which is more suitable for analyzing reality than for clothing it in a vivid description, fails us.

Let us begin with the south-north axis:

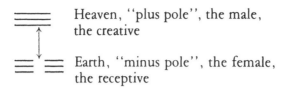

≡ Heaven, "plus pole", the male, the creative

≡ ≡ Earth, "minus pole", the female, the receptive

Designation: Emergence into existence

This prerequisite for all things that are formed and exist in the polar mode is the axis of emergence through polar — in fact yang-yin — unfolding. Here the polar tension between partners which profoundly complement and involve each other is most obviously dominant; here, a symbol of love and marriage is presented as beginning, foundation and — because of its constant renewal — also as the goal of male-female relations, which unites sexual fascination, eros as fascination, love of beauty, existential security and inner harmony. Read the beautiful hexagram of conjugal union, I Ching No. 11 Peace and its opposite No. 12, the disturbance of "marriage", stand still

Here the poles draw apart from each other (tendency to separate). A quotation from the I Ching concerning No. 11

"The Receptive, which moves downward, stands above; the Creative, which moves upward, is below. Hence their influences meet and are in harmony, so that all living things bloom and prosper."

The validity of the axis for *all* domains of nature is realizable here: in the atom as the tension between the negative electron and the positive nucleus; in the animal kingdom as the plus-and-minus helix of the DNA; in the vegetable kingdom in the way all plants rooted in the earth seek the light; in the mental sphere in the polar relation between the immaterial idea and the material appearance — in the Buddhistic terms: emptiness and form —.

In the second horizontal axis we find

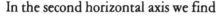

Light	The Abysmal
Fire	Water
Looking	Dim perception

Designation: Becoming Aware

in other words, there is a polar link between two opposite trigrams. The function "perception" links the two together in our interpretation. Let us leave aside here (see above) understanding on a physical plane; on the vegetable plane we have, more clearly than on the first axis, a perception, however vague, of the dark, moist earth on the one hand and, on the other, an active tropism towards the warm, light sky and sun (heliotropism), the whole being intuitable as a unit, which is precisely a "becoming aware", an act of perception; on the "human-mental" plane the parallel is found in intuitive (passive) irrational experience, which C. G. Jung also relates to rational thinking, penetration by understanding.

To this crossed axis,

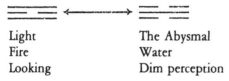

whose validity for all natural creatures should be obvious, we now add the southwest-northeast axis:

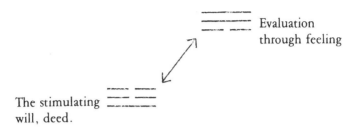

Designation: Effectuation

In this cosmic function, it is still more difficult for us to imagine its existence in the physical and vegetable realm. But in the animal kingdom, the polar relation between the psychic function as an intensive passive acquisition of experience through feeling, and its evaluation (e.g. the many sense modalities of one of our friends in the animal kingdom, the dog) and the thrusting, active life manifestation of the motor, aggressive type provides a rich field of observation for the behavioral scientist, who finds an ever increasing number of points of identity between animal and human behavior! Indeed: we have no difficulty in grasping "feeling and will" as the field of active, operative existence. In hexagrams 42 (Increase = marriage of heaven and earth) and 32 (Duration = marriage) the trigrams are united.

The fourth axis rounds off the orchestra of life and mental images:

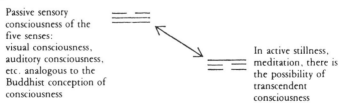

Designation: Becoming conscious

Here it is probably only the passive role, commencing at the most primitive levels of consciousness, elicited and constituted by sensory perception, (witness the mind we can intuit in the dog!) and extending to its highest levels, that is capable of being apprehended to its highest levels, that is capable of being apprehended by us through feeling; the active pole of transcendental consciousness, the efflorescence of humaneness, religion, experience of the divine and metaphysics, is possible only on the human plane. Meditation, active suspension of activity, cessation of the stream of passive sensory perceptions and (in Buddhist terms: thought consciousness as a 6th sense) of uncontrolled, automatic thought make possible the highest level of creation: the wise man, the holy one of the I Ching, who sees through the seeds of fate and conducts himself centrally and in complete harmony with the "law" — and also the luminaries of all great religions: in Christianity in particular, with special stress on personal love imbued with responsibility in imitation of Christ. In hexagrams 41 ("Decrease" with stress on simplification, asceticism?) and 31 ("Wooing", once again a sign of matrimonial harmony) these trigrams are united.

Thus the pattern is completed, the circle of the functions of existence of the polar-linked

Emergence of everything that exists

Perception of light and shade (plants and animals)

Effectuation in sensation and will (animals)

Becoming conscious (human beings) through the integration of the qualities of sensation and consciousness to form consciousness of the *whole*. Thus, applied to the living strands of DNA, our designative pattern provides a means of understanding how, with the emergence of this plus-minus double strand (yang-yin), polar capabilities of *perception, effectuation*, and *becoming conscious* are constituted whose simple phenomenal forms are open for every biologist to study, even if he uses a different terminology (behavioral science, etc.). The fact that it is difficult to answer the question, "What is consciousness?" even in animals, let alone plants, is no argument against the reality of such consciousness.

I Ching As World Formula

While John Kendrew coined the highly appropriate name of "Book of Life" for the genetic code with its 64 signs, I Ching signifies Constancy and Change, or simply: The Book of Changes. This compendium of Chinese natural philosophy has two things to offer: a catalog of 64 hexagrams (= states) which Leibniz mentioned with admiration in 1713 on publishing his binary number system, and at the same time their changes and transformations into another of these 64 states. Thus the I Ching claims to be nothing less than an exhaustive catalog of all the sequent events, processes and developments of Nature for everything that exists as a distinguishable world, everything that evolves from an inconceivable "invisible origin" (Jean Gebser) into a three-dimensional existence and a threefold sequence of time (past, present, and future). For Leibniz, the most important aspect of the I Ching characters was, the parallel between them and his binary system of numbers (dyadic). The sequence of the "hexagrams" in his writings is determined by the binary series of numbers and does not correspond to that of the I Ching (see below). In the above-mentioned meritorious translation of Leibniz from the French (1968) by R. Loosen and F. Vonessen (with an epilogue by Jean Gebser: On the 5000-year history of the binary system of numbers of Fu-Hsi — G. W. Leibniz — Norbert Wiener): "Two Letters on the Binary Number System and the Chinese Philosophy", Leibniz comments in particular on the parallel between the binary system and the I Ching. He also deals in detail with many metaphysical relationships between the I Ching and the Christian philosophy of life in the Occident. Clearly, what he has in mind in this context is a crowning union, in fact, the identity of two such mutually remote worlds through a universal "pansophy", "ars combinatoria" "lingua naturae", indeed, a "scriptura universalis". An enterprise which, after centuries of fragmented research in innumerable special lines, is recognized as legitimate in the works of Einstein, Planck, Schrodinger and Heisenberg. What is involved is finding a world formula, the need for which is now clearer than ever before, and is no longer shunned or indeed scoffed as once it was.

In an essay by Werner Heisenberg: "The Unity of Nature in the time of Alexander von Humboldt and Today"[1], which is at the same time the text of a lecture in memory of Alexander von Humboldt, the author, after a historical review, considers the unity of Nature which, after being lost to view and then repeatedly reappearing in our own day, is examined in the light of modern science. He speaks of its surprising reappearance in the principles governing the smallest particles of matter; he speaks of the surprising emergence of morphology, the science of forms, without the application of which "the behavior of atoms cannot be understood".

"Here one needs the persistence of forms which Bohr postulated with his theory of stationary orbits. In the light of this theory, physics and chemistry appear as a unity, and it can hardly be doubted that sooner or later biology will also be drawn into this unity."

He goes on to say:

"Unity will be restored through the forms underlying all events which are themselves in turn the expression of certain fundamental properties of symmetry in the natural laws . . ."

"What is entailed here are the mathematical transformation properties in completely abstract spaces, discernible only by the mathematician behind the colourful variety of phenomena." After some further thoughts on the "primacy of abstract form . . . in modern biology" with a report on the chemical code of genetic information inscribed on the strands of the DNA molecule, Heisenberg states in a lapidary phrase: "It is our conclusion, from observation, that a few fundamental symmetries and basic forms suffice through their repetition and interaction to create the infinitely complex pattern of observable phenomena." In the I Ching and its symbols, which was found by Fu-Hsi in BC 3000 through an insight into Nature that defies our understanding, and then arranged in its present form by King Wen, there is a world formula of natural philosophy of a kind that might stimulate and enrich Occidental thinking, just as it was profoundly influenced for two thousand years by the much simpler and scientifically unfounded atomic theory of Democritus. There is no doubt that some such half-formulated idea was present in the mind of Leibniz but the synoptic chapter on the I Ching in his great work on Chinese philosophy was never completed; he died while writing it. The work ended in the middle of a sentence — like Bach's last fugue.

Let us now reproduce here the parallel between the binary number system and the I Ching in the version available to Leibniz, the sequence of the hexagrams being determined by the binary row of figures and not the numbering of the I Ching, perfectly acceptable technically, if a working hypothesis is adopted (since the binary series begins with 0, the decadic system counts only up to 63, symbol 63 being the first in the I Ching!). Our best course is to reproduce here the table from the aforementioned book: Two Letters on the Binary Number System and the Chinese Philosophy, Dyadic and IH-King (to adopt Leibniz's spelling). Instead of the Chinese characters for the hexagrams, there follows under E, the translation by Richard Wilhelm.

Binary System and I Ching

A Decadic system
B Binary system*
C Hexagram of the I Ching with number, sign and R. Wilhelm's translation
D Chinese name of hexagrams
E R. Wilhelm's translation

Table according to the book: Leibniz G. W., "Two Letters on the Binary Number System and Chinese Philosophy".

A	B	C	D		E
0	000000	☷☷	2.	K'un	The Receptive
1	000001	☷☳	24.	Fu	Return
2	000010	☷☵	7.	Shih	The Army
3	000011	☷☱	19.	Lin	The Approach
4	000100	☷☶	15.	Ch'ien	Modesty
5	000101	☷☲	36.	Ming I	Darkening of the Light
6	000110	☷☴	46.	Sheng	Pushing Upward
7	000111	☷☰	11.	T'ai	Peace

*Editors note: The binary system is also written 0 and 1 in English.

A	B	C	D	E
8	00L000		16. Yu	Enthusiasm
9	00L00L		51. Chen	The Arousing
10	00L0L0		40. Hsieh	Deliverance
11	00L0LL		54. Kuei Mei	The Marrying Maiden
12	00LL00		62. Hsiao Kuo	Preponderance of the Small
13	00LL0L		55. Feng	Abundance
14	00LLL0		32. Heng	Duration
15	00LLLL		34. Ta Chuang	The Power of the Great
16	0L0000		8. Pi	Holding Together
17	0L000L		3. Chun	Difficulty at the Beginning
18	0L00L0		29. K'an	The Abysmal
19	0L00LL		60. chieh	Limitation
20	0L0L00		39. chien	Obstruction
21	0L0L0L		63. Chi Chi	After completion

64

A	B	C	D		E
22	OLOLLO		48.	Ching	The Well
23	OLOLLL		5.	Hsu	Waiting (Nourishment)
24	OLLOOO		25.	Ts'ui	Gathering Together
25	OLLOOL		17.	Sui	Following
26	OLLOLO		47.	K'un	Oppression
27	OLLOLL		58.	Tui	The Joyous, The Lake
28	OLLLOO		31.	Hsien	Influence
29	OLLLOL		29.	Ko	Revolution
30	OLLLLO		28.	Ta Kuo	Preponderance of the Great
31	OLLLLL		43.	Kuai	Breakthrough
32	L00000		23.	Po	Splitting Apart
33	L0000L		27.	I	The Corners of the Mouth
34	L000L0		4.	Meng	Youthful Folly
35	L000LL		41.	Sun	Decrease

A	B	C	D	E
36	LOOLOO	䷳	52. Ken	Keeping Still
37	LOOLOL	䷕	22. Pi	Grace
38	LOOLLO	䷑	18. Ku	Work on What Has Been Spoiled
39	LOOLLL	䷙	26. Ta Ch'u	The Taming Power of the Great
40	LOLOOO	䷢	35. Chin	Progress
41	LOLOOL	䷔	21. Shih Ho	Biting Through
42	LOLOLO	䷿	64. Wei Chi	Before Completion
43	LOLOLL	䷥	38. K'uei	Opposition
44	LOLLOO	䷭	46. Sheng	Pushing Upward
45	LOLLOL	䷝	30. Li	The Clinging, Fire
46	LOLLLO	䷱	50. Ting	The Cauldron
47	LOLLLL	䷍	14. Ta Yu	Possession in Great Measure
48	LLOOOO	䷓	20. Kuan	Contemplation
49	LLOOOL	䷩	42. I	Increase

A	B	C	D	E
50	LLOOLO		59. Huan	Dispersion
51	LLOOLL		61. Chung Fu	Inner Truth
52	LLOLOO		53. Chien	Development
53	LLOLOL		37. Chia Jen	The Family
54	LLOLLO		57. Sun	The Gentle
55	LLOLLL		9. Hsiao Ch'u	The Taming Power of the Small
56	LLLOOO		12. P'i	Standstill
57	LLLOOL		25. Wu Wang	Innocence
58	LLLOLO		6. Sung	Conflict
59	LLLOLL		10. Lu	Treading
60	LLLLOO		33. Tun	Retreat
61	LLLLOL		13. T'ung Jen	Fellowship with Men
62	LLLLLO		44. Kou	Coming to Meet
63	LLLLLL		1. Chi'en	The Creative

Method of Transcription

It cannot be readily seen from this parallel that, according to the principles and practice of the I Ching, the positive, creative ("plus" pole) principle yang occurs in 2 forms, a resting and an active one (an "excited" state in terms of chemistry), and the negative-female-creative unfolding ("minus" pole) principle yin is also expressible in a resting and active form in double lines.

"old", resting yang   "young", active yang   breaking line becomes
                        with the successive
                        stage "old" yin

"old", resting yin   "young" active yin   uniting line becomes
                        with the successive
                        stage "old" yang

If we were to write out the hexagram in the detail given in the commentary, each of the six positions would have to be described with more accuracy, viz. with an indication whether the line is young or old.

Instead of the "telegram form" the formulation would have to look like this:

6 ――― old yang

5 ――― old yang

4 ――― old yang

3 ― ― young yang

2 ――― old yang

1 ― ― young yang

To be read from the bottom to the top.

This mode of writing is not customary in the I Ching, but to anyone conversant with its laws, it is logical and therefore "permissible" — the later commentaries, for instance, ventured to interpret the hexagram as nuclear trigrams:

1st trigram 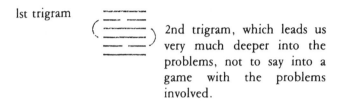 2nd trigram, which leads us very much deeper into the problems, not to say into a game with the problems involved.

The application that follows justifies this digression; the strangeness is due to the newness of the territory we are entering; we shall first have to familiarize ourselves with the mode of writing. For in the following juxtaposition of the genetic code — a field in which the re-emergence of the unity of nature Heisenberg had in mind in his essay, is clearly manifested in the I Ching — the I Ching is not applied as hexagram, but as trigram. (The binary numbers are represented in the customary manner with 0 and 1) E.g.:

0 0 0 0 0 1 would correspond to: old yin
old yin
young yang

Of course, the binary number, previously regarded as unarticulated, now also appears in a rhythmically altered form 00 00 01; in other words, the binary numbers appear in a resting or an "excited" state.

The success attained by using this unusual variant of binary arithmetic in our following investigation, will be its justification — its consistent logic may also prove persuasive to the mathematician. To repeat once again:

Let the basic yang-yin phenomena with their dynamic variants be juxtaposed with their equivalent binary numbers as dyads

$$1\ 1\ \ 1\ 1\ \ 1\ 1\ =\ \overline{\overline{\overline{\underline{\ \ \ \ \ }}}}$$

This mode of writing enables us to bring the genetic code (the reader is reminded of its universality — its four letters, its 64 code words required to describe the 20 amino acids plus 3 "punctuation words") into coincidence with the ancient system of the I Ching with its 4 "letters", its 64 code words, with which, according to the teaching of the I Ching, all sequent events of a physical, psychological, sociological, biological and indeed moral character can be described.

Let U be expressed by ▬▬ ▬▬ or 0
                                                                  0

     C "   "   " ▬▬▬ ▬▬ or 0
                                                                 1

     G "   "   " ▬▬ ▬▬▬ or 1
                                                                 0

     A "   "   " ▬▬▬▬▬ or 1
                                                                 1

Later, (see below, p. 93) this apparently artificial abbreviated form, (from hexagram to trigram) will be fully justified by the fact that, it can be completed to form the full hexagram, when read as a unit with the counterspiral.

$U = \equiv\equiv$ $C = \equiv\equiv$ $G = \equiv\equiv$ $A = \equiv\equiv$

or T in the case of DNA
(Preliminary trial, symbols exchangeable)

	U		C		G		A		
U	0 … 32 … 16 … 48 … 1|2 3|4		4 … 36 … 20 … 52 … 1|2 3|4		8 … 40 … 24 … 56 …		12 … 44 … 28 … 60 …		U C G A
C	1 … 33 … 17 … 49 …		5 … 37 … 21 … 53 …		9 … 41 … 25 … 57 …		13 … 45 … 29 … 61 …		U C G A
G	2 … 34 … 18 … 50 …		6 … 38 … 22 … 54 … 1|2 3|4		10 … 42 … 26 … 58 … 1|2 3|4		14 … 46 … 30 … 62 …		U C G A
A	3 … 35 … 19 … 51 …		7 … 39 … 23 … 55 …		11 … 43 … 27 … 59 …		15 … 47 … 31 … 63 …		U C G A

The I Ching transcribed into genetic code

Here the binary symbols and their decadic equivalents
are shown side by side

0	000000	4	000L00	8	00L000	12	00LL00
16	0L0000	20	0L0L00	24	0LL000	28	0LLL00
32	L00000	36	L00L00	40	L0L000	44	L0LL00
48	LL0000	52	LL0L00	56	LLL000	60	LLLL00
1	00000L	5	000L0L	9	00L00L	13	00LL0L
17	0L000L	21	0L0L0L	25	0LL00L	29	0LLL0L
33	L0000L	37	L00L0L	41	L0L00L	45	L0LL0L
49	LL000L	53	LL0L0L	57	LLL00L	61	LLLL0L
2	0000L0	6	000LL0	10	00L0L0	14	00LLL0
18	0L00L0	22	0L0LL0	26	0LL0L0	30	0LLLL0
34	L000L0	38	L00LL0	42	L0L0L0	46	L0LLL0
50	LL00L0	54	LL0LL0	58	LLL0L0	62	LLLLL0
3	0000LL	7	000LLL	11	00L0LL	15	00LLLL
19	0L00LL	23	0L0LLL	27	0LL0LL	31	0LLLLL
35	L000LL	39	L00LLL	43	L0L0LL	47	L0LLLL
51	LL00LL	55	LL0LLL	59	LLL0LL	63	LLLLLL

Editor: "1" is also written "L". Thus 010100 = 0L0L00.

These two tables show the experiment in which the letters of the genetic code are transcribed in the symbols of the I Ching (which can at the same time be written in the symbols of the binary system of numbers).

Combination of The Genetic Code and The I Ching in a Single Table

In the following table, these five systems have been combined for the first time, and can be examined together:

1. The I Ching, the Book of Changes, (by Fu-Hsi, 5000 years old) with its 64 dynamic states of tension between the opposites of yang and yin.
2. The binary number system, which was conceived by Leibniz and seen by him to be astonishingly similar to its Chinese predecessor, and since been used by Norbert Wiener as the mathematical basis of cybernetics.
3. The genetic code (U-C-A-G) with the 64 triplets). This is the order of the "letters" in the table of the deciphered genetic code (p. 77).
4. (In the margin) The 20 amino acids, the form and content of the whole vegetable and animal kingdoms, programmed by the genetic code.
5. Our decimal number system.

The parallels, and the way in which the two codes fit perfectly into each other, add up to a phenomenon which simply cannot be argued away. All the same, this must be regarded as a provisional experiment. As the arabic numbers in the margin show, this sequence does *not* reveal a mathematical order. In Richard Wilhelm's peerless translation of the I Ching, each of the 64 symbols is supplied with a detailed text and commentaries written over thousands of years. It would go far beyond the compass of this book to attempt to condense still further the meaning of these explanations, since they are already highly compressed, and presented with stenographic terseness. Nevertheless, these descriptions will, in themselves, give the reader a first impression sometimes of static states, sometimes of dynamic tendencies.

However, if we turn the sequence A-G round to give G-A without, so far as I can see, any disturbance necessarily occurring in the arrangement of amino acids, (since, of course, the A-G sequence also seems to be arbitrarily selected), there emerges — all at once — a precise mathematical ordering of the whole, in the sense that the catalog of the amino acids, like that of the codons, appears to be periodically ordered. If the rearrangement I have made in the order is legitimate, it seems likely that interesting and useful inferences might be made, and the transcription into the binary system would be of heuristic value. The sequence is to be read at intervals of four units from left to right (0, 4, 8, 12) and from 16 in the vertical direction (0, 16, 32, 48). In literature to date on the genetic code, which has assumed avalanche proportions, the author has found no reference to a mathematical regularity in the sequence of the codons or their resulting periodic arrangement.

Does our arrangement provide a mathematical explanation of the direction of rotation, the "storeys" of the double helix?

The codes of the I Ching and DNA combined in one table
Editor: "1" is also written "L". Thus 010100 = 0L0L00.

#	code	codon	#	code	codon	#	code	codon	#	code	codon
0 (Phe)	000000	U^{UU}	4 (Ser)	000100	U^{CU}	8 (Cys)	001000	U^{GU}	12 (Tyr)	001100	U^{AU}
16	010000	U^{UC}	20	010100	U^{CC}	24	011000	U^{GC}	28	011100	U^{AC}
32 (Leu)	100000	U^{UG}	36	100100	U^{CG}	40 (Tryp)	101000	U^{GG}	44 (Amber)	101100	U^{AG}
48	110000	U^{UA}	52	110100	U^{CA}	56 (Stop)	111000	U^{GA}	60 (Ochre)	111100	U^{AA}
1 (Leu)	000001	C^{UU}	5 (Pro)	000101	C^{CU}	9 (Arg)	001001	C^{GU}	13 (His)	001101	C^{AU}
17	010001	C^{UC}	21	010101	C^{CC}	25	011001	C^{GC}	29	011101	C^{AC}
33	100001	C^{UG}	37	100101	C^{CG}	41	101001	C^{GG}	45 (GluN)	101101	C^{AG}
49	110001	C^{UA}	53	110101	C^{CA}	57	111001	C^{GA}	61	111101	C^{AA}
2 (Val)	000010	G^{UU}	6 (Ala)	000110	G^{CU}	10 (Gly)	001010	G^{GU}	14 (Asp)	001110	G^{AU}
18	010010	G^{UC}	22	010110	G^{CC}	26	011010	G^{GC}	30	011110	G^{AC}
34	100010	G^{UG}	38	100110	G^{CG}	42	101010	G^{GG}	46 (Glu)	101110	G^{AG}
50	110010	G^{UA}	54	110110	G^{CA}	58	111010	G^{GA}	62	111110	G^{AA}
3 (Ileu)	000011	A^{UU}	7 (Thr)	000111	A^{CU}	11 (Ser)	001011	A^{GU}	15 (AspN)	001111	A^{AU}
19	010011	A^{UC}	23	010111	A^{CC}	27	011011	A^{GC}	31	011111	A^{AC}
35 (Met)	100011	A^{UG}	39	100111	A^{CG}	43 (Arg)	101011	A^{GG}	47 (Lys)	101111	A^{AG}
51	110011	A^{UA}	55	110111	A^{CA}	59	111011	A^{GA}	63	111111	A^{AA}

Transcription of both codes in the binary system
Editor: "1" is also written "L". Thus 010100 = 0L0L00.

Psychological Impediments To An Order of Reality

Reproduced here for the first time, this table shows how these two "books", so remote from each other in origin, can fit together with astonishing neatness. One, a scarcely known compendium of philosophy from ancient China, the other a brilliant piece of modern research with vast implications. Both claim to be universally valid, but whereas the Western mind is immediately ready to accept the truth and reality of the genetic code, it will, as is its habit, flinch away from the truth and reality of a philosophy, to seek refuge in a sceptical and defensive attitude. From the time science was expelled from the bosom of Mother Church, until its coming of age in the 20th century, this inhibition, which can be readily interpreted in Freudian terms as defence, repression, displacement, disavowal, undoing, or even transformation into its opposite, has been characteristic of the relationship between science and metaphysics, philosophy and religion. And perhaps this is also the source of the defensive attitude adopted towards psychoanalysis itself, although this would be the very instrument for exploring the womb of metaphysics; it would have the courage to do so without shame or fear of incest. Yet it is precisely this that is shunned and fearfully avoided; psychoanalysis itself is deliberately dismissed, denied, and suppressed at universities for as long as possible, whereas theology, which seems harmless to science because it no longer makes any scientific claims, is "generously" awarded something like twenty as many chairs. These explanatory remarks on the actual state of affairs at universities is necessary, as a precautionary and salutary measure, to prevent the scientific proof offered here of the identity of a philosophical and a scientific system being automatically rejected and belittled by an immediate "allergic" response.

What is it our Western intellect fears so much that it prefers blindness to sight? This fear must no doubt be seen as reactive, for the pure desire to quest after knowledge was for long centuries (until 1600 AD) inhibited, indeed castrated, by Great Mother Church, because "credo quia absurdum" (I believe although it is absurd) was an unchallenged maxim. This fear of castration, or the greater part of it, is still unconsciously felt in his bones by every scientist. Its recurrence is warded off by the aforementioned defence system: a strict dualistic separation of physics and metaphysics, restriction of research to the material — while the Manichean underground movement of all Christian churches, the divorce of flesh and spirit, is dragged along as an unconscious heritage; hence the denial and failure of scientific research to investigate morals, mind and spirit. Only this schizoid split in the scientific mind itself can have made possible the use of nuclear fission as a proven means of annihilating people who think differently. For since the exodus of science from the church 300 years ago, under the motto "We leave heaven to the angels and the sparrows", Western science has completely and explicitly dispensed with a fixed ideological base, a philosophy requiring no sacrificium intellectus (sacrifice of reason), unless one is prepared to make do with the ideal of an allegedly absolute freedom of research and a vague personal feeling for humanity and order.

Now after two world wars, the atomic bomb, together with growing world pollution and its dangers for man and beast alike, have made the consequences of this lack of system only too clear. The worldwide failure of ecclesiastical institutions might cause anxiety even to those lukewarm "freethinkers" who, secretly, still depend upon the power of the church to establish order. Even the pious and childlike faith of some Nobel prizewinners — carefully segregated from their scientific commitment and deposited as it were in a separate compartment — has failed here completely, along with the equally carefully conserved patriotic, Fascist or Marxistic ideology in the next mental drawer. And yet this abysmal state of affairs is perpetuated and hotly defended. Here again, the appearance of a real order is feared like castration. Even so, the great system of order is already beginning to take shape in men's dreams. In its Christian form in Teilhard de Chardin, in the socialist thinking of Mao attempting to avoid fossilization in permanent revolution, in the initial efforts of the League of Nations and the UN in the political field, in modern art in the uniform rejection of the representational, in science fiction, in a socially and economically integrated world which is uniformly self-controlled by means of giant computers, and indeed, even in the admittedly distorted picture of a world united in sport as seen in the Olympic Games.

Now, is not the I Ching, the standard work of that philosophy which inspired Confucius and enabled him to create a social, moral and religious system that lasted 2000 years, the very book that provides a highly relevant possibility for a new order? Deeply religious — without a personal God! A philosophy whose well-balanced mathematical basis derived from the primary polarity of yang and yin (plus and minus) is acceptable to every scientist without involving a sacrificium intellectus. A system whose scheme of 64 elements anticipated the binary system of Leibniz and Norbert Wiener by five thousand years! A social design with strict emphasis on the primal family (but without violence), parity of "father" and "mother" (in spite of the all-too-human practice found later among the Chinese), free from compulsion and yet bound by reciprocal respect and reverence for the supreme powers (the 8 primal images). A wellbalanced culture which makes possible works of the sublimest beauty in many fields of art, which by virtue of its essential tolerance accepted Buddhism and brought it to its finest flowering. A complete catalog of moral order, of the consequences of appropriate (= good, correct) and inappropriate (= bad, harmful) behaviour — for the Chrstian enviably free from hypocritically and ostentatiously moral pruderies. And a sexuality which is well-ordered because it is honored with reverence is, after all, for the I Ching nothing but the perfected reflection of the primary image of yang and yin. A very similar culture, insofar as polarity is represented in its philosophy and religious art, is the *Tibetan*; however, its profound and prehistorical links with the Chinese cannot be proved in these pages. In numerous depictions we see couples sexually united in yab-yum (= father-mother) position, the "peaceable" and the "angry" gods with their female counterparts, the dakinis, surrounded by aureoles of flame, and also the so-called dyanibuddhas with their female counterparts, both of which have metaphysical significance: e.g. method (male) and knowledge (female) which are "effective" only when united, an example taken from religion and showing the way in which polarity occupies a central place of honor in an original and very ancient culture.

Polarity in The I Ching and The Genetic Code

Looked at closely, the DNA helix created by the union of male and female germ cells presents a picture of extraordinary erotic intensity. In their precisely arranged polar pattern, the purine bases (A-U/C-G) are very precisely combined with the corresponding purine bases of the half of the oppositely turning double helix, like the eternally embracing couple yab-yum of Tibetan mysticism.[1] Seen thus, the world held together in its inmost being by *glutine amoris* (Augustine's "glue of love") and essentially consisting of nothing else would, in fact, be the counterpole of the pure reason of the indifferent but potentially creative world ground, or whatever it may be called. This conception has been presented in full by S. Friedlaender in his "Schopferische Indifferenz"[2] in which a philosophy conceived on precise polar lines shines in lonely perfection among the many dualistic philosophies of the West. The confusion between *dualism* and *polarity* since the time of Heraclitus, which is devastating and pernicious and may well soon destroy us, is illuminatingly described by Jean Gebser in his book on "Dualismus und Polaritat"[3]. Here let us once again stress the authentic nature of polarity and its omnipresence in the DNA code. Seen in terms of both the unbroken sequence of pairs in the DNA plus-and-minus strands, and the precisely complementary codons, sexual consummation appears to be merely a special case of copulation of an infinitely short duration. However, the polarity of the universe, which exists independently of our dualistic interpretation of reality, like the DNA conjugation, the eternally precise polar equilibrium of the "quantities" of all existing things on either side of the "zero", is representative of the equivalence, self-representation, and realization of universal consciousness that transcend all our ideas. The *uni-*

verse is the *object of subjective universal consciousness.* The very few models for such a view provided by European philosophy (because of its dualism) during the last few centuries is clearly shown by Jean Gebser in his book "Dualismus und Polaritat" (Nikolaus von Kues, Paracelsus, Leibniz, Goethe).

How remote the practice of sexual consummation is (and never more so than in the distorted caricature of the sex wave) from the existential, mystical, illuminating possibilities of sexual love, has been shown to Christians in particular by the Zen philosopher Alan Watts in his book "Nature — Man and Woman"[4] and described as a way to satori. Even this existential sexual experience was barred by the predominantly dualistic attitude of the churches and the diabolization of sex for 2000 years. The new integral consciousness which Jean Gebser has revealed, and which transcends by far, the magical and mythical protoforms of earlier ages (shamanism, primitive religion of the Bon in Tibet, etc.) makes it possible to manipulate polar thought consciously, and thus restores the balance that has been lost for so long and with such catastrophic results. We can assist the birth of the New Consciousness — or be our own gravediggers, become our own grave. If the birth goes well, will mankind survive and the new integral consciousness bring to flower new forms of cultural, sexual, religious, political and social existence? Will the infantile age of mankind (psychoanalytically comparable to the behaviour of a three-year-old child in its sadistic-masochistic stage) be followed by a more mature one? (Psychoanalytically: the stage of reality testing: if I soil the earth uninhibitedly, I shall suffocate!)

This is the next step demanded of us — a step which will brook no delay and has already been taken by the vanguard of mankind. In spiritual terms it is the ubiquitous consummation of polarity, just as in ancient China every change was introduced by the imperial announcement of the conceptual content henceforth to be attached to the ideograms. This doctrine of polarity which has proved its efficacy over thousands of years is embodied in the I Ching. To study the book brings inestimable benefit to everyone and it can never be exhausted in a lifetime.

In these pages the author, a country doctor and a profound admirer of the I Ching, presents it to the public in all its significance as an order of reality attested by modern science and confirmed as fact.

And now there is one further aspect to consider: the I Ching in its status as a world formula in the field of biology (and probably also in chemistry and physics).

It must be admitted, in all sobriety, that the West has never had a concept possessing the multiplicity and universality of this one — but it is of Chinese origin and therefore remote from and forbidden to the Western spirit. For what can be more powerful than the prejudices we have mentioned of chauvinism, nationalism, and secularized Christianity!

I Ching The World Code and DNA The Life Code — A Key?

If occidental formulae, scientific laws, structural plans, ideologies, moral schemes, philosophies, and religions do in fact simply point *to something*, provide an intelligible explanation of an existing phenomenon (or non-existent idea which is taught and believed in), and are in the last analysis uncommitted, discussible, accessible to argument, and exchangeable with other quite different schemata (with the appropriate concept adapted to reality representing precisely what we call "exact" natural science), the I Ching appears to have an existence of, for us, almost unimaginable concreteness and vitality which *unites* the *design* and *substrate*, the world object, indissolubly, and indeed seems to be life itself, being at one and the same time metaphysical design and incarnated, vital form which, incidentally, has become legible in the genetic code. And even if, to begin with, we accept this I Ching world formula with its polar symmetries and basic forms, with its claim to represent the actual structure of the intellectual cosmos, merely as a doctrine, a system, as a world code and — as briefly attempted early in this volume (p. 31) — juxtapose it to the genetic code, compare the many parallels attentively, consider the statistically highly significant probability of so many similarities (particularly the universal claim of both systems) being something more than fortuitous, our rational powers, if only because of the relationship set up in our thoughts, begins to circle round the problem and make ever new approaches to it.

The principle of polarity inherent in both systems, the world pole yang-yin on the one hand, the precisely symmetrical plus and minus strand of the DNA on the other, and the very marked congruence of the 64 signs when the two systems are combined, makes tenable the hypothesis that here we have *one* code which pulses through immaterial information and at the same time through the material, but exquisitely delicate programming substrate of all life with its 64 syllables — word that has become flesh. Starting with the scientifically verified DNA code, our mind, seeking understanding and unity, is dimly aware of identity with the world code I Ching. And from the I Ching, accepted "as true", the DNA code acquires the status of meaning and transmitting the law, in other words — the one confirms the other. As we have already stressed, to the scientist who disqualifies his ego from study and thus rules out the subject of scientific research, such an act seems very strange and indeed forbidden and taboo. But anyone who defies this taboo may well be seized by a deep respect for a way of understanding Nature which to us is incomprehensible and inaccessible (and pejoratively referred to as "merely" intuitive, as speculation) yet now fits in extraordinarily well with reality. And with equal ignominy generations of "philosophies" may now be seen as empty structures of thought (Indian: samskaras) — or in modern parlance "paper tigers" — whereas those two decorative Chinese dragons engaged in their eternally undecided contest for the pearl (the world pole of yang and yin is represented in each of our body cells by the plus and minus strands of DNA with its precisely equilibrated 64 static-dynamic fields of force) would seem to acquire the character of reality!

While letting our minds play round this very strange-seeming hypothesis, that a system of natural philosophy thousands of years old is identical with the innermost reality of all life, just as in a computer readout a dancing circling beam of light plays round until the computer identifies figures and letters out of the data presented, let us also approach our theme tentatively and intuitively as in a modern psychological test.

The time has come to examine the world key I Ching more closely:

This sign has a curious look about it, like the germ of a plant, an embryo, perhaps even a ghost, a "bodiless head" freely floating.

And then we turn to the diagram of the DNA spiral with the code written on the "rungs of the ladder", 8 rungs through 360 degree = 1 whole storey of the "staircase". Is there *one* underlying pattern here?

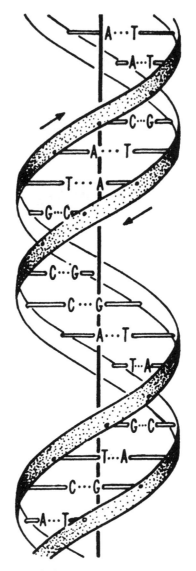

Simplified diagram of the double helix

And then below, combined experimentally, we have the "bodiless head" and the DNA snake: the lower part of the DNA model is freely based on a naturalistic model at the Max Planck Institute, Munich.

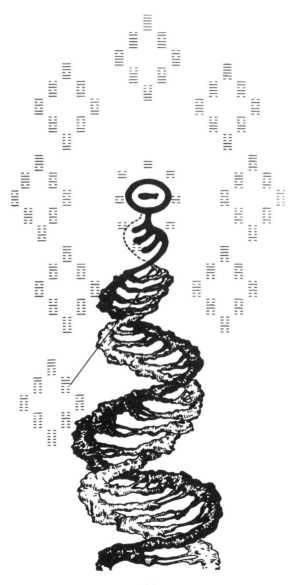

The electron micrographs already available of the double helix reveal an astounding similarity to a perpetually continued I Ching symbol. With its four rungs, the I Ching symbol looks like the head of the DNA "snake". The principle of one turn with four rungs per half storey is the same. In other words, the "ghost" symbol as a drawing is identical with the model confirmed in the electron microscope! Here at last a curious light should begin to dawn in the Western intellect and illuminate the fundamental importance of Chinese philosophy. The written sign[1] which is at the same time the visual image of the DNA double helix is an idea as absurd as it is informative — informative in a sense which is fundamentally different from the previous conception of a "key", for it becomes a key which one can conceive, look at, and possess — and one that fits. What is so astonishing is that the regular light and dark masses of the I Ching symbol complementing each other in a polar pattern suddenly, like a picture puzzle, produce a second, identical but contrarotating I Ching counterimage, just like the contrarotating spiral in the DNA code with the unvarying complementary pairs of bases A-T and C-G.

Any attempt to fit together a written sign and a biological model may seem frivolous and scientifically quite unwarrantable. But this very early ideogram is not a written sign in our sense, but rather the direct expression of a spiritual reality; in other words, it is itself a concentrated image or model of this reality. Was it not Kekule's imaginative picture of a snake of dancing carbon atoms biting its own tail that produced the illuminatng, "fortuitous", idea of the benzene ring and thus the inexhaustible possibilities of organic chemistry? Perhaps the author may be forgiven, then, if he cannot resist the solicitations of imagination.

In the I Ching in all events, the clockwise or reverse-anticlockwise circulatory system is described. It is logical and therefore permissible to make a two-dimensional plan by distributing the eight primal images on a spiral (360°). But what we obtain straight away is the DNA helix with 4 code words each of 3 letters on 1½ turns of the helix. The model is not only similar, but actually the same. A Chinese ideogram, signifying change and formation, and congruous with the DNA helix, exactly as Watson and Crick predicted it would be, it has since proved to be in the electron microscope. As 32 codewords of one descending half of the helix are linked with 32 exactly complementary codewords of the other half, each consisting of 3 transverse rungs, we need 3 whole (360°) storeys to accommodate 8 codewords (codons) or 12 storeys for 32 plus 32 anticodons. The eight-point I Ching star of the trigrams (p. 49) can, as in our "transcription", be written in just such a way, even though it is not yet shown in the I Ching in this form. However, the clockwise direction of reading, which is explicitly distinguished in the I Ching text from an anticlockwise direction, gives us authority to fill in the very same 12-storey double helix with the 32 codons and anticodons of the I Ching.

The two systems are also precisely congruous in respect of the direction of turn and the ascending and descending series of codons and anticodons. The only liberty we have taken is to give the two-dimensional I Ching 8-point star the form of a spiral, i.e. a three-dimensional form. The signs "would in realith have to be drawn as a spiral rising in space"[2]. At the same time it becomes clear that, with the arm of the helix counterrotating as a mirror-image, the I Ching symbol that appears to have been arbitrarily read as a trigram, instead of a hexagram, (p 72) is now in fact completed to form a hexagram, and the parallel, indeed the identity, is still more specific and convincing.

I Ching ideogram continued and completed with its complementary half.

Old China grasped this complex of intellect, spirit and body as a unit, and realized it on a scale that is inconceivable for us — let us hope that, following this unexpected emergence of Chinese philosophy in all its potency, our Western minds will be able to close a schizoid gap in our intellectual life. Our continued existence depends upon it. True, the genius of the West discovered the genetic code, but compared with the total conception of natural philosophy offered here, the sole possession of the code is on a par with the statement of the blind man who, feeling the tail of an elephant, described the elephant as being "shaped like a worm" (Indian legend). Below we shall try to feel the whole "elephant" and describe it. But before we do so, we can no longer postpose a digression in the I Ching as a book of oracles.

Freedom and Programme in The I Ching

It cannot be kept from the reader who has followed so far that, the I Ching has been handed down to us primarily as a book of wisdom, and a compendium of natural philosophy only since the reign of King Wen — even though thousands of Chinese scholars have always thought of it, commented on it, and added to it as such — for in the earliest times it was used to recognize cosmic situations and their future development ("oracles"). By asking questions of Nature, and through constant reference to the totality of natural conditions, it became possible to appraise this general situation and its subsequent development in conformity with natural laws, in the form of a forecast consisting precisely of these 64 hexagrams and their 6 secondary meanings (or none, some, or all of them) and mutations into another of these states.

Without wishing to rationalize this possibility, which has hitherto seemed incomprehensible and incredible, it is perhaps easier for us than earlier generations to understand it because of the knowledge we now have of computers, or "thinking machines", of which Leibniz made an intensive study as far back as the 18th century. With one difference: whereas computer results are based on as many *individual data* as possible in order to produce a correct forecast, the I Ching, having its own inherent conception of reality, starts from this *general situation* and returns a wise predictive judgment and reply when the questioner, the inquiring subject, comes into contact with this totality through a process which may seem very strange to us. While not wishing to go in for trite rationalization, the modern reader, being familiar with the idea of programming sequences of multivariate events and with the great importance of statistics and probability calculus, may, in a way that was previously impossible, acquire an interest in — and indeed an understanding for — the importance of the I Ching as a source of prediction and wisdom based not on the invariably incomplete sum of *single data*, but on the *whole*. The central secret of the I Ching, namely the possession of this "whole", is, of course, in no way profaned by this.

In the new field of futurology, science itself attempts:

1. the prediction of the probable course of technical and its own development,
2. the avoidance of unwanted developments,
3. the optimal programming of a specific development.

The previously scarcely suspected unity of science, and the further development of this science, are here made manifest. From the very outset there was no doubt in the Chinese mind — and indeed no way of conceiving otherwise — that the genesis of the world from the primary poles must of necessity constitute a datum with variable sequences of events, developments, and "fate". Hence, of course, the name "The Book of the Constant (= natural philosophy) and Changes (= naturally necessary development). Now, the fact that form, structure and all characteristics of living creatures are determined by the programming of the life process through the genetic code could, without strain, be defined and interpreted as a programming of their fate. The DNA of a jellyfish contains a fate different from that of an oak tree; that of a canary different from that of a tenor — and yet, one day, the similarity or indeed the identity of long segments of DNA concerned with the voice and the ability to sing might be demonstrated just as the relatedness of all living creatures has been demonstrated in respect, for example, of a quite specific genealogy of the respiratory enzyme, which already enables a species to be allotted a specific place in the pedigree of life. The sum total of the living conditions of a creature, its appropriate locomotion in an optimally suited habitat, even indeed, its life-long social behaviour, is certainly constituted as a datum through the genetic code, however unusual the term "fate" may be for this in science.

Rats behave like rats all their lives, and human beings — like human beings. To what extent we still carry rat-like elements of DNA in our behaviour might become a fateful question for humanity. Will "hawks" or "doves" obtain the upper hand in the DNA fate of the Americans? But to return to our subject: the I Ching as a statement on the naturally necessary development of given "states". Let us recapitulate once again for those who are less well versed in these matters: The 64 states, possible tensions between "plus" and "minus", yang and yin, are defined and transcribed in the book in extremely concrete and lapidary words. For thousands of years commentators have given these great precision. Each of the six lines of the 64 hexagrams has, moreover, a precise and special signification. In the case of the hexagrams which have a fundamentally adverse significance, the 6 lines often contain more favorable and consoling tendencies; in the case of very harmonious signs, the six special significations often warn of a tendency to decay. The "good", happiness, *must* be shattered — the "hole", the abyss, is filled up and the way to freedom is made smooth. If one or more special significations are "active", in the "excited" state, one hexagram alone allows 26 possible combinations. Because of the tendency of "excited" lines to be transformed into a new state

(6 = → ← becomes ———, 9 = ——⊖—— becomes __ __)

the whole hexagram is altered and is turned into another one of the 64 hexagrams. This, then, intimates a temporal sequence, a programmed forecast in the presence of "excitation", lability.

To such an extent, indeed, that the hexagram ䷀ which indicates the highest possible degree of power or energy can tip over into ䷁ indicative of the purest "absorptive" passive receptivity. All this can be readily understood in terms of electrical processes such as interference, modulated frequencies, and even switching functions in a computer (4096 possibilities). Indeed, in modern terms, the genetic code might be described as the logical circuit of a computer with 64 switching units.

Because of the virtually infinite number of combinations and — importantly — the infinite series of endless rearrangements of code word sequences, 64 hexagrams are perfectly adequate, despite their apparent simplicity, for describing all forms of life. And the system of 64 hexagrams in which the variants of fate are coded in the I Ching is adequate for all imaginable human, social and private destinies seem from inside and in terms of the configuration of fate. This configuration, of course, may be characterized by quite different places, environments and attendant historical circumstances. The psychologist knows very well that a "handful" of mixed drives — compounded of the drives of love and aggression (the latter also the death drive?), or in other words, of two basic drives (yang and yin?) — suffice to analyse the workings of the mind in health and disease. The guidance supplied by this precise scheme of 64 elements is more copious — and more exact. To the European, no doubt, this unity of sex, social role, ethics and the external course of fate must seem strange. But the "skeleton" of the I Ching programming, the sequences of fate, may be seen as fitting together precisely with the sequences of amino acids, as programmed by the genetic code, to form a single entity: that is the *hypothesis* advanced here.

I Ching, Law of Chance

In our first introduction to the world of the Tao, we made known to the reader the two poles of yang and yin, the four phases of power, the eightfold structure of mind-spirit-matter, and the 64 dynamic developments of fate with 6 steps each; but thee is one important point still to mention. In the operation with the yarrow stalks[1], or else with coins or dice, there is a "fortuitous" three-part combination of the remainders of the divided stalks or dice which are placed together as the final result. In this connection 3 counts as a "yang particle" ($= +$), 2 as a "yin particle" ($= -$) and the operation, repeated six times, serves as a means of examining the cosmic chance situation.

Change	Sign	Sum	Abbreviation
2 2 2	▬ ▬	$6 = - - -$	→←
2 2 3	▬▬▬	$7 = - - +$	▬▬▬
2 3 3	▬ ▬	$8 = - + +$	▬ ▬
3 3 3	▬▬▬	$9 = + + +$	▬⊖▬

We know already that these are our 2 stable and 2 unstable conditions, the 4 "letters". We might express the idea this way: In the innermost nucleus of the actually indivisible yang-yin poles, there are still the minutest building blocks which are, however, to be regarded as "unreal" *before* the sum of three particles. For example the hexagram is then constructed from these triple-divided precursors of the _____ and _ _ which, in an unabbreviated form, would actually have to be written as a symbol with 12 parts (see above). The author cannon resist the temptation here of referring to the **quark model** of baryons of quark spin +, which is a product of the SU_3 theory (Gell-Mann) and embodies a concept of the innermost structure of the smallest particles hitherto regarded as indivisible; a model which uses particle charges in three parts, known as quarks, whose existence, however, has not to the best of my knowledge been proved and, it is the fervent hope of their "discoverer" Gell-Mann, never will be! All the same the elementary particles (mass O!) are "thought" to be built up out of still smaller elementary particles with charges ± ⅓ or ± ⅔ of the charge of the electron.

»Quark«	Name		el. Ladung
o o ●	n	(Neutron)	0
● ● o	p	(Proton)	+1
o o o	Δ −	(Delta minus)	−1
● o o	Δ 0	(Delta Zero)	0
● ● o	Δ +	(Delta plus)	+1
● ● ●	Δ + +	(Delta Double plus)	+2

The fact that out of this SU_3 theory it is possible also to develop a SU_6 and an SU_{12} algebra seems to be a very remarkable parallel to the triplet hexagrams and the 12-part system (see above). However, it is the strange world of the hypothetical quarks that the curious results of modern physics are at home: reversal of time, simultaneity of contradictory states, etc., which have shaken the very foundations of classical physics, together with the dualistic philosophy of Aristotle. I now want to juxtapose this model and the astonishing parallels of these two tables — systems of 3, 6 and 12 elements with microparts whose existence and character are indeterminate in time and space — with that function of the I Ching whose results are as difficult for the scientific mind to digest as the results of the SU_3 theory I have intimated, namely, the ability of the I Ching to read the future from fortuitous events.

In contrast to the difficult quest for the ultimate constituents of the electron, this experimental investigation of laws of chance which are unknown to us, is open to anyone who cares to apply the technique which the I Ching hands down to us from early Chinese times, viz. to question and analyze the 64 states and their 6-stage meanings in the I Ching. With the above collation of SU_3 theories and I Ching manipulations as a basis, we might state the following, giving in each case the version of modern physics in brackets: Since for the eager student of the I Ching there is only one world (*informational magnitudes, origination of matter from pairs of electrons*) unfolding in a polar mode out of a state of maximum potency undefinable in space and time (*energy from the "big bang"*) with the triplets of the yang-yin manifestation (*SU_3 theory as a comparison, quarks, vide supra*) corresponding to the transitional field from Tao to yang-yin (*energy into matter*), it is possible, pursuant to the I Ching law of further development into 2 ⎯⎯⎯→ 4 ⎯⎯⎯→ 8 ⎯⎯⎯→ 64 invariably polar (+ / −) states (*atomic structure in eightfold shells, energy pattern of electrons, a maximum of $2_n{}^2$ electrons possible in each shell, see diagram on page 125, one uniform principle as far as the transuranian elements*), to predict rule-governed consequences of these states; in other words, the qualitas occulta of what is born in this moment of time (C. G. Jung) becomes readable through the I Ching[2]. (The exact prediction of the omega elementary particle of the baryons with the aid of the SU_3 group model was a great success). Are we justified in supposing that the same law that orders the minute particles of matter and allows them to be predicted is also active in the origination of the DNA life code?

Chance and Necessity in DNA, Surrealism and I Ching

If the active-constant "pairedness" welling up from Tao appears to us as merest chance-and not as quanta-in an ordered 64-fold variation of the world complex of matter, mind and spirit, this is due, when considered in the light of the I Ching, to our dualistic prejudice, to the illusion of dichotomy between the ego and unintellectual and unspiritual nature. If we conceive our person in the present moment to be a *head*, and the millions of the extremely minute phases of the past, which are of a very specific character because they are constituted by a sequence of very specific "correct" and "incorrect" decisions, to be the long *body* of the serpent of destiny, then the I Ching shows us the next step of fate in the course of Tao.

Every present moment has one quantum of freedom (one in 64). *After* the decision freely taken, the binding law of cause and effect becomes operative; we have to bear the consequences, and the "body" of the serpent of destiny grows longer by one more piece. The "head" is only free within limits, namely those imposed by the appendage of the past determined as it is by local, social, and peristatic factors and forming a rigid "evil", bound, or vitally illuminated totality. This, then, is roughly the picture of chance (fr. Latin "cadentia" = falling, especially of dice) we find in the I Ching: Everything that is chance is subject to the code of law in its 8- or 64-fold structure with personal freedom in the present moment. Chance is cosmic action! Chance is the eternal process of creation structured in accordance with law. And, seen like this, as Monod stresses, the DNA of all living creatures is in fact likewise a product of chance — all plants and animals embodying innumerable series of chance events, with special diversions, aberrations, variants and invariants, and blind alleys, ranging from the virus to the immense personal freedom of man to recognize this law and to live with dignity in conformity with it. Those people who know only statistical chance are, of course, condemned simply to "fearful seeking in an icy, forlorn world, a paroxysm of fear"[1]

Monod believes he must expect man

"to awake out of his millennial dream and to recognize his total solitariness, his radically alien nature. He now knows that he has his place like a gypsy on the edge of the universe which is dead to his music and indifferent to his hopes, sufferings, and crimes".

How different things are for the man who is centered on the center of the eight spiritual possibilities, centered by a life of attentive listening for his aberrations through error! Here we need not hesitate to apply Buddha's Eightfold Path of right living: right view, right thought, right speech, right action, right livelihood, right effort, right mindfulness, right concentration: *this old man sits serenely and peacefully in the center* (instead of at the verge of the universe), profoundly at peace, free from all fears, no longer suffering, with a deep sense of security; *he himself represents the eternal law*. And, in the event of aberration, the natural law of the I Ching responds hypersensitively to false hopes, self-created suffering and crime, with automatic admonition and correction. (No indifference here!)

I Ching and chance in the DNA parallel also provide a solution to the following problems: In statistical terms the "chance" creation of life, defined as the creation of self-producing units, seems to be highly improbable; even assuming the presence of the first of such units, life would have advanced only as far as, say, the bacterium and would not be conceivable without a "driving demon."[2] In our account of the I Ching doctrine of chance with its 64 states, stable and propulsive on the one hand and inhibitory on the other, the "demon" is represented as follows: The tendency to propulsion is dominant in the following 43 hexagrams (1, 2, 7, 8, 10, 11, 13, 14, 15, 16, 17, 19, 20, 21, 22, 24, 25, 27, 28, 30, 31, 32, 34, 35, 37, 40, 42, 43, 44, 45, 46, 48, 49, 50, 51, 53, 54, 55, 57, 58, 61, 62, 63); the tendency to inhibition and decay in 21 hexagrams (*3, 4, 5, 6, 9, 12, 18, 23, 26, 29, 33, 36, 38, 39, 41, 47, 52, 56?, 59, 60, 64*).

This interpretation can be no more than a rough approximation. In cases of doubt, a hexagram was allotted to inhibition. There is therefore a ratio of 2:1! I have subsequently found that this private interpretation of mine is confirmed by the conclusion reached by R. Wilhelm in I Ching, Book II, p. 312, section 5, where he shows that, when the yarrow stalks are counted in every possible way, the result is 6912 yang lines and 4608 yin lines; that is, yang outweighs the negative pole yin in precisely the ratio of 3:2. This corresponds to 11520 "mutations" of the 64 hexagrams (R. Wilhelm). Thus, in the dynamic world principle of I Ching, 3 states out of 5 are active and propulsive, and 2 out of 5 are retarding and inhibitory. An optimistic philosophy, if you will, and one which is given relevance and reality by its coincidence with the DNA. Every doctor knows from experience: Nature heals, the doctor only lends aid. But — man is free to destroy this Nature...

Assuming that we accept the natural philosophy of the I Ching, i.e. that the dynamic function of this doctrine of hexagrams is operative throughout Nature, then the fact that out of the 64 states 2/3 are active-propulsive and only 1/3 negative-inhibitory does not suggest a law of slow decay, but rather a very marked **accelerating** factor (= demon!)[3]. It would be enough not to condemn this purposiveness as "animistic", but rather to adopt it, with success, as a hitherto unknown principle of great heuristic value in explaining the unexpectedly rapid and precise development of the preliminary stages of DNA, into such highly complex representatives of life.

Having confronted the I Ching doctrine of chance with scientific theory, let us make another attempt to understand and appreciate chance in a new way: this time in Surrealism, as it is presented in Christian Kellerer's book "Objet trouve, Surrealismus, Zen".[4] Kellerer believes that a successful work of Surrealistic art affords a real possibility of bridging the gap between statistical and psychological chance. The object found by chance (objet trouve), once it has been divested of its links with everyday life, acquires an enlightening meaning in the eye of the beholder.

However, as Kellerer goes on to say, in the citizen (and the scientist?) who seeks to exorcise "with the holy water of his conventions" anything that casts a shadow over his prosperity and high living standard as "crazy" and "outrageous", this act of perception elicits a mixture of humor and dread, dread being understood as chthonic and elementary fear. A dread which also clearly finds expression in Monod's book (see above). On experiencing a case of elemental chance, the observer, who has been displaced from the center of his experience, finds that the unity between personal being and object has been restored with a shock and a sensation of dread. The experience may be accompanied by amusement and heightened consciousness. Kellerer finds interesting parallels between this incipient enlightenment and the attainment of oneness through Surrealism — admittedly evanescent and fragmentary as a rule — and the doctrine and practice of Zen observed for centuries in China and Japan, where time and again there are reports of how enlightenment is found through a paradoxical chance (through a stone bouncing off a bamboo, or a blow from the master; indeed, in Buddha himself, through a sudden, "chance", view of the rising morning star). "The sack (of the ego) is burst", or "the pitcher is broken", are typical accounts of enlightenment.

For our purposes, we may assume that here "chance" becomes clearly apparent in a sense in which it may be understood by modern, rationalistic man in his frustration. At the same time, its operation is demonstrated in the overcoming of modern man's predicament by the attainment of true momentary (Surrealism) or lasting (Zen) enlightenment. For both Jean Gebser and Christian Kellerer, what is ultimately important is the replacement of the dualistic consciousness by a more comprehensive supreme consciousness, in which polar realities are presented to experience in what is necessarily a differentiated polar form; every DNA which has come into being by chance, and which we encounter as plant, animal, or human being, can then become an illuminating objet trouve, "Every day — good day", as Zen puts it.

Then chance is seen as in Christ's saying (Matthew 10, 29) about the sparrows: "One of them shall not fall on the ground without your Father". Certainly, it would be of interest to examine the problem of chance and necessity from the viewpoint of other religions: chance as seen through the eyes of Christianity, Islam and Hasidism, but this would go far beyond the compass of this book. All the same, in a television discussion on Monod, Manfred Eigen warned against accepting scientific objectivity as the sole criterion, "then we might lose things" (he mentioned charity and mercy) "simply because we cannot include all parameters within the scope of our objectivity".

Every possible parameter is contained within the eight faculties of the spirit, with all-embracing objectivity. Scientific objectivity (ONE absolutized parameter of the EIGHT of the I Ching) which does not take into account the mental structure of the observer (e.g. a depressive character neurosis, a severe unconscious castration complex with existential anxiety) is a pure illusion — as everyone should know since Freud — and it is the absolute duty of every scientist to eliminate such factors which disturb and distort his vision.

We may be quite sure from the outset that, in this sense, the new consciousness will be of an integral character and thus will also be based on a polar combination of equilibrated forces of *intellect AND intuition*, of *feeling AND will*, on the gift of *meditation AND* the most delicate *sensuous perception*, also on a nucleus of the male-female person in indissoluble unity.[5] It is the task of all of us to start the preliminary work here and now.

The Practice of Prediction in The I Ching

And now we come to the nub of the matter, which the Western mind finds hard to accept. First of all, let us recapitulate in brief, the actual practice of the I Ching. The questioner, anxious for specific and concrete details of his fate, performs a ritual of actions and manipulations with the aid of stalks (yarrow stalks), coins, even dice, in which he seeks at the same time to meditate and concentrate his mind on his question, so that the manipulations or the throw are performed by a questioning subject. A hexagram is built up of the four "letters" (written from the bottom upwards) obtained by six manipulations (see p. 101) either without (= 7, 8) or with accentuated "moving" lines (= 6, 9). The hexagram is looked up in the book and the questioner, with his question in the forefront of his mind, automatically and inevitably brings the information into association with the question. His political, private, or sociological question is clothed in "flesh and blood" by the questioner in response, say, to answer No. 23 = *Po* — "splitting apart", and the predicted "splitting apart" is interpreted as political (e.g. war), private (e.g. divorce), or social (e.g. revolution). Question and answer undergo a remarkable unification or "incarnation". That, then, is the process in brief.

Needless to say, further details could be given of the actual correspondence of prediction and reality, on the prediction coming true. C. G. Jung gives an example of such an experiment conducted by Richard Wilhelm, the unexcelled translator of the I Ching, in the Psychological Club, in which the oracle was confirmed by subsequent events[1]. In many cases the author was able to assure himself of the predictive power of the I Ching, which is admittedly incomprehensible by subsequent events with what might be called statistically significant frequency. In spite of the persistence with which he asked questions concerning the projected construction of a house (official approval of the plan had been obtained), he obtained predominantly (80%) negative predictions and never one of the affirmative hexagrams. The plan as it was — at that time was thwarted by "external" circumstances, and the house was never built. Frequently, the answer hits the nub of the problem with an unsparing accuracy which none of the other hexagrams in the book could have supplied.

Many of the I Ching predictions have the character of a parting of the ways, a finely shaded but neutral formulation intermediate between the path of freedom and the path of fate. We may put it like this: if you conduct yourself in conformity with what is cosmically the optimum, then the following will happen in fulfillment of the laws of Nature. — It is entirely open to the questioner *not* to take the optimum step advocated by millennial wisdom, and then the consequences of this step will also follow in obedience to natural laws. Here the I Ching and the Indian doctrine of karma comes close together. From the moment the points are set one way or another, the "train" continues along its inevitable route. There is freedom *at the* moment of decision. There is *one* optimal path, a "being free in the most loving must" (Nietzsche), but there are *many* doleful windings and turnings in the labyrinth. Within a complex rule-governed pattern, which so far escapes our comprehension but includes the questioner, his hand, his problem, and the I Ching "computer", the I Ching oracle shows the optimum path which runs "in Tao".

Only 15 years ago it would have seemed absolutely incredible to assert that all the processes of life were recorded in an unchanging code on a long double strand. Today, extensive research suggests that there are good grounds for believing that experiences and memory can be shown to be located on the same DNA substrate. And it does not appear to be too far a step from memory to programming and fate. The question is this: Does the 6-foot long DNA strand which is coiled up in each of our cells and contains the approx. 5,000 million "words" of instructions for constructing the body, also embody the program for behavior, adaptation, social structure, in a word, our fate which appears to be so extremely individual, personal and private only because of the ego illusion, and, analagously to our memory, also incorporate our life memory and program — nay — contain them simultaneously and automatically as a datum in its own constitution? Is not the classification we make: genetic information, unfolding in a time sequence (past, present, and future), information about fate not the illusion, the colored reflection of *a spectrum* (incidentally, the eight colors suffice for all the colorfulness of the world and all painters and paintings in the world) into which our differentiating intellect refracts the light of life? I believe that this is precisely what the I Ching is saying: this is a spiritual place.

One thing is certain: looking back over a whole life, we see it as unalterable and strictly determined: this is how it was and not otherwise. The sum total of billions of tiny steps in life, minutes of fate, years of decision have crystallized round a nucleus which is nevertheless highly characteristic and apparently non-fortuitous when seen in retrospect — crystallized like crystals on a thread dipped in a hypertonic solution of salt. Our fate might be seen as forming a unit with the billions of code words of our DNA strand! And in the nightmares of the geneticists in which they manipulate the DNA and produce superhumans and subhumans to order, they view fate as manipulatable, too, and capable of being given a very special cast . . . How refreshing to see the I Ching at the opposite pole as the high, unshakable suprahuman order of natural law which, for

5000 years, was the innermost principle in Chinese society, family life, civilization and art. And, confirmed by the findings of the geneticists, it might be consciously accepted as the ruling principle for thousands of years to come.

Let us consider the "oracle" once more as the "program": Our conception of chance and free will lacks scientific precision. For even what is apparently chance is governed by natural law, the necessary result of countless physical, chemical and also spiritual forces. So long as science excludes spiritual orders and forces, its view of total reality will be mistaken, however accurately its conception of matter. In the I Ching there is full respect for physics-metaphysics, which is still regarded as inseparable, for moral law and personal aberration, however undeveloped the conception of chemical and physical details may be. *My* unique question and *my* unique hand movements, which are the upshot of countless microscopic determinants that could be accurately described (but never are because of sloth and alleged scientific irrelevance) and which occur in my existential, unrepeatable moment and can never be reproduced, are in reality strictly determined. In this context, as in all others, strict natural law prevails. Thus disregard of this determination — and that is what we call our belief in "chance" — can be shown to lack precision.

If the questioner is now brought close to the center of his existence by the method of meditation (and meditation is not mystical but a method of mental application describable in exact psychological and physiological terms), then the casting of the coins, the stalks or the dice takes place *from this center* and can be read in the pattern of the 64 hexagrams because, in accordance with I Ching doctrine, everything that comes to pass is expressed in this code, and therefore my question, my problem and my "chances" can be read there in conformity with the law. The practical experience of those versed in the I Ching is an extremely lively proof to anyone who takes it seriously. All the same, there are many gaps in this attempt to describe the "oracle" for, wherever we look, there is no basic scientific research into the microscopic physical processes of "chance" and next to none into the efficacy of spiritual forces. And yet it should be obvious enough to us how a few fantastic ideas, i.e. intellectual forces (racialism, "Jews are our undoing"), have brought ruin and destruction to millions, reduced cities to glowing ashes, led to Hiroshima — all highly concrete consequences.

The question: What are negative, "demonic" forces? and what is the path these forces take to the bitterest and most keenly felt results can be fruitfully asked in scientific terms, only if an attempt is made to discern "forces" as effective reality. But our science does not do this — any more than medical science teaches "health" or "healing", That is something the doctor must discover for himself — and curiously enough he actually does so; only in this way does he become a doctor. And just as the doctor must discover for himself — and curiously enough he actually does so; And just as the doctor is born only when subjectivity is introduced into medicine — the doctor who represents what medicine really is in his own person (Paracelsus: "As the strength of man lies in the woman, so the strength of the physician lies in the patient") — so the introduction of the observer as subject has seemed to be an indispensable element of modern physics ever since Einstein[2]. Modern physics, of course, knows precious little about the structure of this subject. It is still a very long way from giving serious consideration to the scientific research on a broad front achieved by Freud's psychoanalysis and still less to the infinitely wider ranging psychoanalysis of Buddha, or the I Ching doctrine of the 4 polar passive-active faculties of the spirit (page 44), at least three of which are unknown to our classical psychology.

The I Ching and The Genetic Code in the 5-Stage Pattern of Meditation

If we try to adduce proof of the connection between, or indeed the identicality of, the I Ching principle of world creation and order on the one hand, and the most recently discovered programming code of all living creatures, DNA, on the other, so as to obtain an immediate insight — which alone is conducive to further development — and make this available to the reader, we again find a very ancient system immediately at hand. In view of the growing interest shown in Zen Buddhism and Eastern mysticisms in general by leading Western minds such as Aldous Huxley and Hermann Hesse in literature, Herrigel in philosophy, and P. Lassalle S. J. and Lama Angarika Govinda in religion, we may venture to assume a sufficiently wide acquaintance among the public with meditation and its uses. All the same, the system of meditation used here is perhaps known only to the readers of Gundert's "Bi-Yan-Lu"[1] and Dumoulin[2]. It is the system of meditation of Master Dung-Shan (807 to 869).

It is the purpose of this system of meditation to enable the meditator to obtain insight into the fundamental unity of two apparently divergent fields, viz. in Buddhism the unity of the Absolute and the Relative. This is achieved with the aid of *five stages* or localities.

Dumoulin makes explicit reference to this system as an expression of the I Ching philosophy. With this as our warrant, and with our given theme, viz. philosophy (= "absolute") and biology (= "relative"), we will try to give the reader an insight into the connections between the I Ching world formula and the universal life code of DNA, which will be all the more profound because it is obtained through his own activity.

Dumoulin (p. 122 ff.) refers to a subsequent change in the order of the five stages, which was therefore not regarded as fixed for all time.

The original system is:

1. The Absolute in the Relative.
2. The Relative in the Absolute.
3. The Absolute alone.
4. The Relative alone.
5. Absolute and Relative in polar union.

After careful reflection we selected the order with the Absolute in the first place, and the Relative in the second place. The other places then follow logically.

Thus, in terms of our two themes, we contemplate:

Stage 1 — the I Ching — the "right", the "master" (the absolute, transcendental, metaphysical, "beyond time and the world"),

Stage 2 — the genetic code = the "one-sided" or "crooked", the "servant" (the materially formed, physics, the "here and now"),

Stage 3 — the way of the "right" into the one-sided" (= the "master" goes to the "servant"), of the spiritual principle "into" nature.

Stage 4 — the way "back" from the "one-sided" to the "right" (= the "servant" goes to the "master"), from nature "back" to natural law,

Stage 5 — the "right" and the "one-sided", "crooked" as an identical unit (= "master" and "servant"), what is "beyond time and the world" as the "here and now", the world key *is* the world. Our dualistic view of the world is an illusion, there never was an "into" and "back", the apparent chaos is the cosmos.

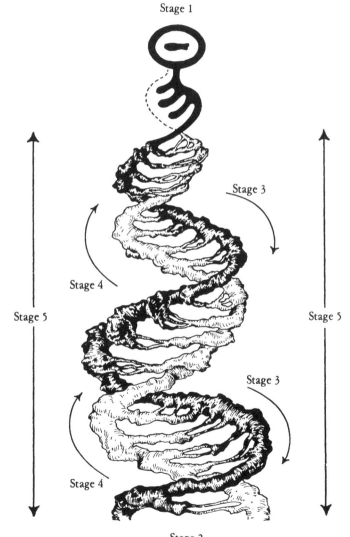

EXECUTION
STAGE 1:

This is represented by the I Ching as cosmogony, as "world formula", as elemental order, as the system of the eight primal forces with which we are already acquainted (page 44), and their 6 times 64-fold states of transition with which all conceivable changes and developments can be described. They may correspond to Heisenberg's conception of "symmetries and basic forms which suffice to bring about the infinitely complicated pattern of observed phenomena by repetition and combined action" (p. 55 loc. cit.). Cosmogony, natural philosophy and also the system of physical, mental and biological order constitute data in the I Ching's doctrine of reality. The reader will remember the curious natural order of the four axes of Emergence, Perception, Effectuation, and Becoming conscious, which can be applied to the whole biological field as an exact universal psychology, far more comprehensive and true to reality than many a Western model.

This part of the I Ching involving world view, cosmogony, psychology, characterized by its piety without God, its perfect "electrical" plus-minus structure, seems to resemble atomic designs and be acceptable even to a scientist; whereas the second part of the I Ching doctrine with its 64 states, each with its 6 variants, seems strange. It floated — and still does — in a void as something alien and unbelievable; an abstruse "equivocation", an oracle book catalog (described as equivocation even by that outstanding sinologist Gundert). The sophisticated ethics and family system of the I Ching might be described as a third part. Another main section, the fourth, is the dynamic, programming tendency such as has been discussed above. This system, measured against all other natural philosophies, religions, and scientific systems, has seemed unique. Indeed, until now, extremely peculiar and even "irrational", meeting so far with no great response. In the words of the Foreword, "an erratic block, a prehistoric find, which refuses to fit into a well-ordered landscape". It existed in detachment, unproven for those uninterested in it, abstract, abstruse and, in addition, disguised as a "book of oracles", slandered as such, and present to the minds of very few.

STAGE 2: THE DNA CODE

What is entirely new is the knowledge of the ultrastructure of the DNA double helix. Only just visible in the electron microscope, the double helix with its plus and minus strands, which embodies in 64 code words, each made up of 3 out of 4 possible letters and forms, the astonishingly unified programming of plant, animal and human life in all its colorful variety. In suspense, as it were, in the human consciousness as a highly curious novelty, it was recognized after a short spell of astonishment as a promising field of research. Already it is being proclaimed that the *atomic* age will be followed by the *biological* age, the age of biological manipulation in which the creation of horror-film monsters, even the possibility in theory of reconstructing giant dinosaurs, seems to be the "acme of possibilities"![3]

This is the "earthbound", material, aimless, and very specific result of modern science which has so far never dared, never ventured, to focus its attention on the spiritual law or immaterial forces: namely, to consider the fact, vast in its implications, that by means of a *script*, by means of a purposive intelligence of an impersonal character, millions of different structural plans are written, and in such incredible plenitude that the plan is contained whole in each cell. Does anyone seriously imagine that a Jumbo Jet or an Apollo rocket can come about by a statistical chance? Who designs the Apollo rocket of Nature —

man? Here, obviously enough, the "rest" of the elephant (= stage 1) is missing from the "elephant's tail" in the design of the genetic code. It is fascinating to realize that the whole of living nature consists of libraries, book-like plans, that have become flesh; the scaffolding is at the same time the building itself.

STAGE 3

From the way of the "right" to the "one-sided". The abstract, transmundane, transcendental law or — as we may express it in modern terms — the law consisting of information changed after "the big bang" from an ineffable state, the "invisible origin" (Jean Gebser), into matter, into the inconceivable quanta of positive and negative voltage, into temporal qualities of past-present-future (and according to Heisenberg there are in the atomic sphere "certain processes which seem to proceed in a reverse direction compared with their causal sequence"), into the atomic nucleus and electron shells (8 electrons per shell) all the way down the periodic scale to the massiveness of the heavy metals.

This may be read, interpreted and followed as a logical path leading from immaterial (informational) quality, structure, to thick solid lumps of matter. From the hypothetical quarks, postulated in the SU_3 theory of Gell-Mann, who interestingly enough, made successful use of a Buddhist system of the "Eightfold Way" to arrange the swarms of elementary particles (just as we are here using the Buddhist pattern of 5 stages of meditation) to form photons (mass 0), gravitons, electrons, myons (neutrinos), mesons, and baryons. We may quote the work of E. H. Graul and H. W. Franke[4].

"This brings up the question of how the variety and color of the world have come about. May one perhaps speculate that

the ultimate really elementary constituent of our world has <u>one</u> <u>single property</u> *which occurs in only* <u>two</u> *states. — Then only* <u>one</u> *information bit is ncessary for its description . . . It may be that the fundamental constituents of our world are informational in character. Then our world would be reducible to information".*

In an article by Prof. Fritz Popp, Department of Physics of the University of Munich, entitled "Research on the Fringes of Physics"[5] we read

". . . early in the thirties it was found that, on colliding with charged particles, light quanta of sufficient energy may disappear and be replaced by pairs of electrons, one positive and one negative, without any change in the particle with which it collided. Since colliding electrons in turn produce light quanta again, which may once more be reconverted into pairs of electrons, electrons increase very quickly in numbers by the continuous production of pairs as long as there is sufficient energy available. With the extremely high energies of cosmic rays, a single quantum of light, which has itself no rest mass, may give rise to a trillion pairs of electrons with a weight roughly equivalent to that of a billion hydrogen atoms. Matter, then, is not something that remains immune to all changes . . .

Further, more and more evidence has accumulated to confirm that there is interdependence in the transformation of elementary particles. <u>Each particle contains all the others in posse</u>. *In other words, the durability of the world is not guaranteed by matter, but by the law that governs the transformations. This law therefore constitutes a question of fundamental importance".*

"It is not to our pictures of the world that we must look for support, but rather to the order of reality, which we can never

BASIC STATES AND ELECTRON ARRAGEMENTS

Z	K 1s	L 2s 2p	M 3s 3p 3d	N 4s 4p 4d 4f	O 5s 5p 5d 5f	P 6s 6p 6d	Q 7s
1 H	1						
2 He	2						
3 Li	2	1					
4 Be	2	2					
5 B	2	2 1					
6 C	2	2 2					
7 N	2	2 3					
8 O	2	2 4					
9 F	2	2 5					
10 Ne	2	2 6					
11 Na	2	2 6	1				
12 Mg	2	2 6	2				
13 Al	2	2 6	2 1				
14 Si	2	2 6	2 2				
15 P	2	2 6	2 3				
16 S	2	2 6	2 4				
17 Cl	2	2 6	2 5				
18 Ar	2	2 6	2 6				
19 K	2	2 6	2 6	1			
20 Ca	2	2 6	2 6	2			
21 Sc	2	2 6	2 6 1	2			
22 Ti	2	2 6	2 6 2	2			
23 V	2	2 6	2 6 3	2			
24 Cr	2	2 6	2 6 5	1			
25 Mn	2	2 6	2 6 5	2			
26 Fe	2	2 6	2 6 6	2			
27 Co	2	2 6	2 6 7	2			
28 Ni	2	2 6	2 6 8	2			
29 Cu	2	2 6	2 6 10	1			
30 Zn	2	2 6	2 6 10	2			
31 Ga	2	2 6	2 6 10	2 1			
32 Ge	2	2 6	2 6 10	2 2			
33 As	2	2 6	2 6 10	2 3			
34 Se	2	2 6	2 6 10	2 4			
35 Br	2	2 6	2 6 10	2 5			
36 Kr	2	2 6	2 6 10	2 6			

Z	K	L		M			N				O				P				Q
	1s	2s	2p	3s	3p	3d	4s	4p	4d	4f	5s	5p	5d	5f	6s	6p	6d		7s
37 Rb	2	2	6	2	6	10	2	6			1								
38 Sr	2	2	6	2	6	10	2	6			2								
39 Y	2	2	6	2	6	10	2	6	1		2								
40 Zr	2	2	6	2	6	10	2	6	2		2								
41 Nb	2	2	6	2	6	10	2	6	4		1								
42 Mo	2	2	6	2	6	10	2	6	5		1								
43 Tc	2	2	6	2	6	10	2	6	6		1								
44 Ru	2	2	6	2	6	10	2	6	7		2								
45 Rh	2	2	6	2	6	10	2	6	8		1								
46 Pd	2	2	6	2	6	10	2	6	10										
47 Ag	2	2	6	2	6	10	2	6	10		1								
48 Cd	2	2	6	2	6	10	2	6	10		2								
49 In	2	2	6	2	6	10	2	6	10		2	1							
50 Sn	2	2	6	2	6	10	2	6	10		2	2							
51 Sb	2	2	6	2	6	10	2	6	10		2	3							
52 Te	2	2	6	2	6	10	2	6	10		2	4							
53 J	2	2	6	2	6	10	2	6	10		2	5							
54 Xe	2	2	6	2	6	10	2	6	10		2	6							
55 Cs	2	2	6	2	6	10	2	6	10		2	6			1				
56 Ba	2	2	6	2	6	10	2	6	10		2	6			2				
57 La	2	2	6	2	6	10	2	6	10		2	6	1		2				
58 Ce	2	2	6	2	6	10	2	6	10		2	6			2	?			
59 Pr	2	2	6	2	6	10	2	6	10		2	6			2				
60 Nd	2	2	6	2	6	10	2	6	10		2	6			2				
61 Pm	2	2	6	2	6	10	2	6	10		2	6			2	?			
62 Sm	2	2	6	2	6	10	2	6	10		2	6			2				
63 Eu	2	2	6	2	6	10	2	6	10		2	6			2				
64 Gd	2	2	6	2	6	10	2	6	10		2	6	1		2				
65 Tb	2	2	6	2	6	10	2	6	10		2	6	1		2				
66 Dy	2	2	6	2	6	10	2	6	10		2	6			2				
67 Ho	2	2	6	2	6	10	2	6	10		2	6			2				
68 Er	2	2	6	2	6	10	2	6	10		2	6			2				
69 Tm	2	2	6	2	6	10	2	6	10		2	6			2				
70 Yb	2	2	6	2	6	10	2	6	10		2	6			2				
71 Lu	2	2	6	2	6	10	2	6	10		2	6	1		2				
72 Hf	2	2	6	2	6	10	2	6	10		2	6	2		2				
73 Ta	2	2	6	2	6	10	2	6	10		2	6	3		2				
74 W	2	2	6	2	6	10	2	6	10		2	6	4		2				
75 Re	2	2	6	2	6	10	2	6	10		2	6	5		2				
76 Os	2	2	6	2	6	10	2	6	10		2	6	6		2				
77 Ir	2	2	6	2	6	10	2	6	10		2	6	7		2	?			

Z	K	L		M			N			O				P			Q
	1s	2s	2p	3s	3p	3d	4s	4p	4d	4f	5s	5p	5d	5f	6s	6p	6d	7s
78 Pt	2	2	6	2	6	10	2	6	10		2	6	9		1	?		
79 Au	2	2	6	2	6	10	2	6	10		2	6	10		1			
80 Hg	2	2	6	2	6	10	2	6	10		2	6	10		2			
81 Ti	2	2	6	2	6	10	2	6	10		2	6	10		2	1		
82 Pb	2	2	6	2	6	10	2	6	10		2	6	10		2	2		
83 Bi	2	2	6	2	6	10	2	6	10		2	6	10		2	3		
84 Po	2	2	6	2	6	10	2	6	10		2	6	10		2	4		
85 At	2	2	6	2	6	10	2	6	10		2	6	10		2	5		
86 Rn	2	2	6	2	6	10	2	6	10		2	6	10		2	6		
87 Fr	2	2	6	2	6	10	2	6	10		2	6	10		2	6		1
88 Ra	2	2	6	2	6	10	2	6	10		2	6	10		2	6		2
89 Ac	2	2	6	2	6	10	2	6	10		2	6	10		2	6	1	2 ?
90 Th	2	2	6	2	6	10	2	6	10		2	6	10		2	6	2	2 ?
91 Pa	2	2	6	2	6	10	2	6	10		2	6	10		2	6	3	2 ?
92 U	2	2	6	2	6	10	2	6	10		2	6	10		2	6	4	2 ?
93 Np	2	2	6	2	6	10	2	6	10		2	6	10	4	2	6	1	2 ?
94 Pu	2	2	6	2	6	10	2	6	10		2	6	10	6	2	6		2 ?
95 Am	2	2	6	2	6	10	2	6	10		2	6	10	7	2	6		2
96 Cm	2	2	6	2	6	10	2	6	10		2	6	10	7	2	6	1	2 ?
97 Bk	2	2	6	2	6	10	2	6	10		2	6	10	8	2	6	1	2 ?
98 Cf	2	2	6	2	6	10	2	6	10		2	6	10	10	2	6		2 ?
99 Es	2	2	6	2	6	10	2	6	10		2	6	10	11	2	6		2 ?
100 Fm	2	2	6	2	6	10	2	6	10		2	6	10	12	2	6		2 ?
101 Md	2	2	6	2	6	10	2	6	10		2	6	10	13	2	6		2 ?
102 No	2	2	6	2	6	10	2	6	10		2	6	10	14	2	6		2 ?
103 Lw	2	2	6	2	6	10	2	6	10		2	6	10	14	2	6	1	2 ?

These tables show the uniform principle upon which the elements are built up from hydrogen to the transuranian metals and, according to which, the electron shell of each is to be regarded as arising from the preceding one by the addition of one more electron.

The whole periodic system can be conceived as being built up from hydrogen (H) by the successive increase of the nuclear charge by one unit (= +) and the addition of a further electron per element.

Each period of the system corresponds to an electron shell saturated with 8 electrons. The ordinal number (extreme left = 0) corresponds to the number of positive charges of the nucleus and thus also of the electrons in the atom shell.

In what appear to be deviations (e.g. 2 + 8 + 8 + 18) sometimes only part shells are completed (in which the total, orbital and intrinsic angular impulse = 0). Otherwise "normal" 2 + 6 = 8 electrons = saturated shell, with the outermost shell always being built up to 6, the outermost shell (1, or 2, or 3 ff) determining chemical activity while the saturated shell is very stable and "quiet".

(It would be reasonable to make a comparison with the 1-6 "active" lines of the I Ching and the "unstressed" lines; the 6th line would be followed by the next symbol).

It is a system which up to iron (Fe, 0 26) is impressive in its uniformity and, since it is not very familiar to the layman, it is reproduced here. Consisting of *one* positive and *one* negative charge matter gathers like an avalanche right down to the heaviest and densest elements.

comprehend fully with pictures, although we may draw ever closer to it".

Now the I Ching *is* the law by which the 64 states change into each of the other 64 states; the sentence quoted above might have been taken from the I Ching. —

In a text by Dr. Siegbert Hummel under the title "Polarity in Chinese Philosophy"[6] we read:

"The Tao posits one, one posits (at the same time as itself) two, two posits (at the same time as itself) three. This three posits all creatures".

In terms of the I Ching: "Programming" by the transcendental, immaterial, 64-element computer scheme of the 64 conceivable states of tension effectively determines the direction from immaterial BEING as energy to material existence. Here we have a precise "system of reality" such as Fritz Popp calls for.

We have an approximate idea of how things then proceeded on our earth and probably on many other earths. Out of the seething primal soup came organic substances, simple amino acids formed "by chance" (?), a very primitive metabolism arose, and then the great throw: chain molecules, precursors of DNA, precursors of viruses appeared, processes became increasingly permanent and consistent as they were coded on the strands to form programs, and duplication of these programs yielded propagation! Unicellular then multicellular organism, then the whole tree of vegetable and animal life. Calculation has shown one thing: that through the operation of pure chance, development would scarcely have advanced as far as the virus; in other words, there is in each living creature a headlong dynamic impulse which intelligently avoids dead ends. How, whence, who, why? Each answer found gives rise to new questions. After countless deviations and diversions came the phylum of mammals leading to man: pongids, ramapithecus, australopithecus — homo habilis — man.

And now rational man has discovered the key of all life, the genetic code, the uniform set of building blocks with its unending variations; he seeks to read himself. Was this the aim of the Big Bang?

We recapitulate in meditative terms: primal polarity, "exploding" Tao, and a 64-element structure-code-principle in infinite variation suffice for the origination of the elements, their combinations into suns and planets, plants, animals, and men. The path from immaterial, informational entities can be investigated in an already logically complete and unbroken sequence all the way to man himself with his ability to recognize, contemplate and reconstruct his whole "path".

STAGE 4:

From the path of the "one-sided" to the "right". After matter had been fully formed into its elements, plants, and animals, there seemed to be nothing transcendental left at all. In the world seen as hyle, as material, in the sorrow of the creature, in the cold and slimy reptilian world of the third day of Creation, in its eating and being eaten, spirit and law are lost to sight and memory; the inconceivable energy of the origin of the galaxies with a temperature of many millions of degrees, this surge from out of the realm of pure energy, seems to have come to a dead stop. In spite of the appearance of rational man, there are great abysses of suffering, error, confusion, war, disease, murder and torture. And out of this gulf of fixity, uncertainty, suffering and pathology, after many millions of years of "descent", there opens on the fringes of the human mind the pure miracle of the way "beyond time and the world": in the ineffable smile of Buddha and his psychoanalysis of suffering that surpasses all later psychology, *deliverance from suffering* becomes visible, teachable, and capable of being experienced. Almost at the same time as the human spirit was flowering in the Mediterranean area — in the sculpture of classical Greece, in the imperishable and incorruptible words of Socrates, in the great new beginning of personal love and responsibility Christ brought into the world — there appeared the new path "beyond time and the world" which, it cannot be denied, is beset at all times by the danger of new error (hence the great fear of "heresy" or wrong doctrine fought with fire and the sword). Buddhism lays explicit stress on the singular possibility of human existence, of acquiring knowledge, of complete deliverance; he therefore places human existence even above the "worlds of gods".

Anonymous, prehistoric nature religions (shamanism, totemism), the doctrine of "mana", the cult of the dead reaching its apogee in Ancient Egypt, might be cited as preliminary stages of the way "beyond time and the world", from the "servant" to the "master". And in prehistoric China, 3000 years before our era, the I Ching was formulated by Fu-Hsi, the sole testimony among all the world systems of all philosophers that points to a pattern of 64 states as a code and universal law. Five thousand years ago the Chinese found the exact law, the transcendent, in all its plenitude. Besides these and other travellers on the path from the here and now to the Beyond and besides the founders of the great world religions, an untold number of unnamed mystics have passed along the "path", have sought the way to enlightenment and described it in remarkably similar terms. There are, for example, the formulations of Meister Eckehart which have astonished Japanese Zen-Buddhologists because of the similarity of their vocabulary to the teachings of the classical Zen masters. And it would be absolutely right to see the epistemology of modern science as forming part of this path, proceeding from infinitely diverse material forms to the uniform law, to the unformed principle (which has so far, unfortunately, been oversimplified as an abstraction, an unreal "as — if"). In the formulations quoted above where "information" is regarded as the essence of the universe, one side of the "right" is seen with precisely the same exactitude as in the I Ching but seen, as it were, *monocularly*! In the I Ching the vision is *stereoscopic*, binocular, and embraces not only the structure but also the potency, the dynamic of the immaterial as a universal law, as universal and eternally valid morality, and at the same time the structural program of the world. A world formula of explosive energy, not merely a written character!

Let us once again recall the striking similarity, indeed identicality of the image and reality, i.e. of I Ching and the genetic code, brought home to us by the two pictures

I Ching symbol   DNA symbol

How much the Western intellect would benefit from examining such powers reverently and circumspectly, from respecting the reality of spiritual forces! In the I Ching and its doctrine of the 8 powers and their mingling in the states of inhibition and promotion (sometimes fortunate, sometimes tragic, sometimes physical and sometimes psychological) in a clear and logical computer circuit diagram encompassing the transformation of these 64 symbols, and in a noble polar system venerating the animality of sex as a central phenomenon, the West could acquire (without the dogma or the sacrifice of common sense inherent in almost all religions except Buddhism) in addition to its sharp eye for details, an equally sharp eye for the universal, the generally valid and the profoundly human — and see the world *whole* and in *polar* terms not *dualistically*, distorted and alienated. Meditatively we repeat: The genetic code with the 64 code words now known opens up an avenue of thought: From "here and now", from the incarnated mathematics of the genetic code (see page 78), there is a holistic and true philosophic understanding which leads above and beyond the "as-if-character" of most philosophies and the irrational dogmas and unreal structures of belief of most religions, direct to the transcendent "way back" — into sight there comes a path found by mystics through groping intuition in the metaphysical quest without this knowledge, but often also missed because of the innumerable dead ends and wrong turnings. A path of meditation of supreme precision, an exact metaphysics, an undivided unity of nature and spirit.

Perhaps it will not have gone unnoticed that the disjunction between these stages is not clearly defined, that they merge one into the other — because, after all, they represent an artificial intellectual dichotomy of a single entity of being.

STAGE 5:

The identity of the "right" and the "one-sided", of the "transcendent" and the "here and now", of the universal key I Ching and the DNA code of life.

Identity can be consummated and experienced only in the realization that both codes are *one* code — it is all too easy for insight to focus upon the identity of the two cornerstones: the world design of the I Ching and the life program alphabet each with its 64 code words, and yet — how infinitely "insubstantial" these two cornerstones of nature appear to be. Millennia have passed and taken barely any notice of the I Ching. There are millions of people totally ignorant of the structure of their DNA life thread. In spite of the whole "library" present in each cell, the mass of the DNA strands of the whole of humanity would fill only a walnut shell — a delicacy that defies the imagination. And yet they are matter. How infinitely more delicate is the stage preceding these strands, the spiritual design of the 64 words in both cases. Everything else, the overwhelming plenitude of the world, has become "flesh" round this inconceivably delicate thread, all the destiny, all the givenness of the world, can be expressed through the 8 powers, the 64 symbols. We have the esoteric knowledge of the I Ching initiate, his divinatory power based on his insight into the mathematically prescribed and not "fortuitous" nature of all things, and we have the esoteric knowledge of the future geneticist, which one day will be able to predict the deliberate and planned development of given DNA — highly embarrassing and undesirable and indeed dangerous though they may both legitimately seem to many. And yet the destiny of the future world might depend on the contact, the union, brought about between the overwhelmingly radiant order and ethic of the I Ching and the undirected and amorally rootless knowledge of the possibilities of genetics with its suicidal blind alleys.

Let us try once again to bring together the two "ends" of the "elephant" (if we may recall the crude metaphor) of which we blind philosophers on the one hand, and blind scientists on the other hand are vaguely aware and know intuitively to be a unit. The pattern of 64 items in both cases, its 4 letters of which 3 are used, the perfect way in which (as shown above for the first time) the two codes may be written into each other, the universal claim of the I Ching philosophy on the one hand (cosmogony, programming of all sequences of development) and the geneticist on the other hand (all life built up, maintained and propagated in accordance with this 64-symbol code). With the dividing line between animate and inanimate nature becoming ever more fluid, the question arises: How large is the portion of animate nature compared with the "inanimate"? The I Ching philosopher had no knowledge of the high-velocity movements of each atom, and the geneticist is still ignorant of the continuation of the 64-symbol code into the prebiological realms of nature and indeed to the origins of the world. From the "invisible origin" (Jean Gebser) before time and space were divided, before spirit and matter were divided, down to homo sapiens and his history, the world code becomes discernible as the inmost framework of the whole of nature, as a universal script and language, as the Esperanto of Nature.

Without THE LAW as the omnipresent framework of the whole of nature, man does, of course, seem to the molecular biologist Jacques Monod to be the "chance hit of Nature" — there *is* an enormous sense of forlorness and sadness if the MEANING (Chinese: Tao) is not known. In the I Ching of Fu-Hsi, as in Laotse's book of Tao, this basic law of the universe is present "in persona". Its acknowledgment by modern science on the basis of its agreement, its unification, with the indubitable discovery of the genetic code might be compared, insofar as this sadness, rejection and forlorness, and indeed the probable self-destruction of fallen man is concerned (the law would not be affected or impaired by this destruction since it is not identical with the human) to the good fortune of a Starfighter pilot who is about to crash and at the last moment finds the lever of his ejector seat.

Meditation on the 5th Stage

Instead of persevering to the end with this meditation on the 5th stage (unity of the I Ching world system and the natural system of DNA), instead of elaborating on the experience (and result) to be expected, I venture to recount an incident from my own student days. One dismal autumn day, I found my way, tired, hungry and in a hopelessly confused state of mind, through a dilapidated wall into a wood and wandered hither and thither without a path through the wilderness, over a meadow, came upon a canal with a narrow, ramshackle bridge, stood on the middle of it and raised my eyes to a picture of breath-taking beauty: I was standing on the longitudinal axis (the canal) of a huge park, from which majestic avenues of strictly ordered centennial trees, glorious in all the colors of autumn, opened out symmetrically. The setting sun, breaking through at this moment, gave a fiery light to the trees and sank below the horizon exactly in the axis of an infinitely long "sun street" of reflections. I turned round — and saw at the other end of the canal — the Royal Castle with its fountains, flights of steps, and a hundred incandescent windows which looked as if they were lit from within, but were actually illuminated by the reflecting sun which, only on this evening of 23 September, set exactly at the western end of the axis! The moment grew into eternity and brought pattern and order to my poor, bewildered existence.[7]

The reader will have no difficulty in making his own comparisons: Sun = I Ching; the "sun street" and canal = DNA code of 64 elements from "here" to "there"; castle and park = the whole of the animate cosmos; "enlightened" man = the subject of all objects; from the supposedly gloomy and chaotic dereliction of the "chance hit man" (Monod) there is vouchsafed through a universal key of natural philosophy of a polar integrating and not dualistic character, through a *formula of reality* (see Popp), a view of reality in the perspective of enlightenment — and the earth becomes paradise. Seriously: because for 2000 years Western thinking has been dualistic and has lacked the polarity it is so vitally important to maintain (just as the strands of DNA are always exactly combined in a precisely polar manner) the whole world has been seen awry. He who subscribes one-sidedly only to the good, recognizes only a God the Father (and not God the Mother), and uses fire and sword without inhibition, since sex, decried and disparaged, is provided with a safety valve in this way, loses control of the spaceship earth and is incapable of using the right course-correcting control systems — for control is always polar in its function, since it involve a continuous and hypersensitive reaction to polar opposites.

In the bibliography attached, the reader will find the right guides: The comprehension and practice of the I Ching doctrine of polarity with its venerable patina of thousands of years comes very easily on using the translation of Richard Wilhelm.

"Schopferishe Indifferenz"[8] by S. Friedlander, probably to be found now only in large libraries, describes a highly practical application of philosophical polarity. Everyone is set the task of replacing the one-sided dualistic spiritual forces of thinking and the will, and if need be of sensation, by true functional pairs which bear a polar relation to each other (p. 50) — perception, effectuation, becoming conscious in their active and passive aspects — and of being finely sensitive to disturbances of balance occurring in conjugated primal polar forces and of correcting them actively.

This is the concluding medical prescription and advice which may help the reader to give practical application to polarity.

In fact, it would bring immense benefit to have a theory of the world acknowledged as binding, i.e. scientifically recognized and generally valid, which polarizes universal indifference and presents a clearly ordered world with a cosmogony of informational data, differentiation of the elements from hydrogen to the transuranic elements, biogenesis from the pre-stages of metabolism to hemoglobin, from simple chain molecules to DNA (now already largely proved) — a continuous process, that is from the first information bit to man himself. In the place of innumerable systems, world views, hypotheses, and partial worlds we should have at least the intimation of a very first approximation to the integration of all these systems. We recall once again the hope expressed by Leibniz in connection with Chinese philosophy that Occidental science and theology might be united with Chinese philosophy in a "pansophy", "scriptura universalis", "lingua naturae" (= language, script, book of Nature) — and this is precisely what would no doubt have happened had a mind of his stature known of the discovery of the genetic code and given it his assent with instant enthusiasm. It is highly probable that he would have immediately realized that the 64-symbol system of the I Ching and the code form ONE system, ONE law. Just as certainly as the reaction of universal minds (and perhaps every general practitioner must have a grain of this universality) will achieve this 5th stage, namely the unification of the I Ching world formula and the genetic code, so will, with equal certainty, minds of the same cast as Wagner, Faust's amanuensis, be negative in their reaction. Based as they are on fixed dualistic notions, they are scarcely capable of recognizing and using such a master key which might bring the specialist's bunch of keys into disrepute. They will continue to reject the establishment of such a relationship as unscientific and undesirable.

For the "whole man" (non-schizoid but rooted in polarity) this 5th stage would, of course, be a profound confirmation, a felicitous re-unification of the sundered and schizoid worlds of faith on the one hand and science on the other. Whether, as a member of a church, he follows the broad pathways of faith or whether, as a philosopher or mystic, he takes the "direct way", the Middle Path, "free from contradictions", the transcendental would again become an inviolable good for all because it would have again become integrated scientificially (just as science conceived as Tao or meaning is in its turn rooted in the "right" of the transcendental) and no longer to be purchased by a sacrificium intellectus.

For even if, viewed in terms of chance, all living creatures are probative designs of a drive of immense energy (Freud's "id", taken over by Groddeck[9] "Das Buch vom Es"), it is possible only with man "the chance hit" as a starting point to go the way of meditation backwards and beyond time and the world, the way of all philosophers which, indeed, goes far beyond Freud's superego and is demonstrated time and again in the I Ching together with the fundamental law, the mathematic frameword of Nature. For the student of Far Eastern tradition there is no doubt that this way has been travelled to its goal, which is release, nirvana, liberation from dualistic opposites, SATORI, by multitudes through the precisely taught methods of meditation, of Yoga, and especially of Zen Buddhism. Perhaps the way in which the I Ching has been accredited by its identity with the genetic code will correct the European's superficial and unthinking dismissal of these Eastern experiences as unscientific and therefore untrue. The knowledge may dawn that the exclusion of facts of mind and spirit (e.g. the transitional field between my wish to move my finger and the movement of my finger) was unscientific and unforgivable mischief.

It is precisely this transitional area which is the real domain of the I Ching. Adjacent to the borderland in which Heisenberg apparently found time moving backwards in atomic processes,[10] appears this window in the wall of our previous conception of the universe. And this wall is the duality man posited 2000 years ago in place of a polarity filling the whole of Nature, indeed a Nature which consists of nothing but polar "quanta". The I Ching like the genetic code, as true textbooks of polarity, might open up for us a new stereoscopic instead of the former dualistic-monocular vision. A vision of a huge energy-laden world of the minutest spatiotemporal quanta, in which Aristotle's "either-or" is of only limited application, whereas the polar "both-and" regains validity in conscious consummation (Jean Gebser, "Dualismus und Polaritat"). Our survival depends upon the consummation demanded.

Summary of All Reflections

Realizing how defective any attempt must be to present, in a clearly thought out form, a discovery which is so difficult to grasp because of its "overwhelming" nature, (Jean Gebser), we shall now give a summary of the contents and the results of this work.

The author begs indulgence for formal errors and for formulations which may seem inaccurate to the specialist. The important thing is to heed the message.

1. The *similarities* in seven points between the I Ching and the DNA code which were published in 1969.

2. A strking similarity was reported between a system of natural philosophy with a code of 64 symbols and the genetic code, which also has a system of 64 elements.

3. Not only *similarity* but also identity of the two codes and the I Ching as a world formula on the basis of the binary numerical system.

4. This *similarity* would be reduced to an unexceptional occurrence of the binary numerical system in the two systems, were it not for the universal claims made for both systems. A disquisition on the two codes and on the I Ching as a world formula, that is to say, as the more comprehensive system which that is to say, as the more comprehensive system which "emerges" in the DNA code as a special case. *A priority claim is herewith made for the I Ching!*

5. The transcription of the DNA code into binary numbers and the I Ching symbols (made easier for the reader by the corresponding arabic figures) does not reveal a mathematical order.

The reversal of the genetic letters A-G to G-A discloses an exact mathematic order, or rather, this order demands that the reversal be made! We must await an expert judgment whether we do not also have here a kind of periodic order of the amino acids; it also remains to be seen whether this order represents a natural law.

The transcription also allows the "punctuation" codons UAA, UGA, GUG to be looked up in the "Book of Changes" — as a kind of check on postulated identity by reference to a concrete example:

UAA = STOP = ䷠ = "Retreat" (No. 33, I Ching)

UGA = STOP = ䷋ = "Standstill" (No. 12, I Ching) i.e. the "meaning" of the codon:

"Stop", termination of a genetic code sequence, is synonymous with the meanings of "Retreat" and "Standstill". The codon:

GUG = START = ䷃ = "Youthful Folly" (No. 4, I Ching)

meaning the youthful nature of the situation and transition to a new one!

The codon UGG, according to my transcription No. 40 — I Ching No. 35 — "Progress", corresponds to the essential amino acid tryptophan. In the I Ching this is one of the most progressive signs, and tryptophan occupies a key position in metabolism. In the research work of F.A. Popp it figures as the donor of *biophotons* (!), i.e. the ultraweak but measurable luminescence of the DNA, which is released more particularly on the death of living cells. In acute puerperal depressions there is a significant tryptophan deficiency and its administration is also therapeutically useful in other depressions. Just as the countersign No. 5 — I Ching No. 36 — "Darkening of the Light" is implicit in the I Ching sign "Progress" so the absence of tryptophan would indicate depression.

Thus the check we made by reference to an example does seem to show a remarkable coincidence, a synonymity. Our conclusion is that there is a further *similarity* in meaning between the DNA code and the I Ching symbol and a principle of order incidentally revealed in the genetic code with the aid of the I Ching.

6. *Similarity* of the basic graphic symbols. One of the parameters which science with its claims to objectivity officially derides but nevertheless diligently uses[1] is creative imagination, which circles indefatigably round a problem until it yields its secret.

The systematic application of this instrument reveals another *similarity* between the two systems which is visually impressive.

a) The ideogram for the I Ching, if it is continued in writing, has the same shape as the DNA helix, which has been photographed in the electron microscope.

b) The clockwise turn of the helix is a mirror image of the anticlockwise spiral.

c) Both impress the observer by the fact that they are sequences which occur, or can be thought out, at great or infinite length — to the extent of billions of double-helix storeys in the case of DNA.

7. *Similarity* with regard to *variance* and *invariance*. Monod believes the special characteristic of DNA to reside in stable invariance and free variance. In translation the title of the I Ching reads: STABILITY and CHANGE.

These *similarities* led to an attempt being made to superimpose the I Ching symbol and the DNA helix. It was possible to write on 24 double storeys not only all the DNA code words needed for the synthesis of the body's proteins, but also intellectual and spiritual structures in an entirely new psychosomatics of the future, by means of the eightfold structural elements of mind and spirit and the 64-fold temporal deployment in behavior, development, "fate" and character which, according to the doctrine of the I Ching, are perfectly adequate for the purpose. At the same time it is possible to extract from this alphabet, in its aspect as both DNA and I Ching, sequences which may be short and primitive or of great complexity. Such possibilities make the vast potential scope of the "discovery" clear: e.g. an "objective" psychology (mesomerism: codon —— "psychon").

8. *Similarity* between the I Ching doctrine of chance and modern theories of physics, such as the SU_3 theory, which can be applied to the ultrastructure of DNA (see diagram for the structure of the I Ching elements and the quarks!). In its method of investigating the emergence of the existing from the non-existing, the I Ching textbook of chance uses the minutest tripartite units with a positive or negative "charge" to construct a line element which only then "appears". The I Ching doctrine of chance also enables statements to be made relating to the chance origination of DNA, plants, animals, and man himself which are similar to those relating to the chance structure of the elementary particles. Further similarities: SU_3 algebra with SU_6, SU_{12} variants, triplet and hexagram variants of the I Ching with 12 elements.

All the same the I Ching recognizes the statistical chance of objective science as only *one* of 8 possibilities. There are precise variants of chance on the basis of intuition, feeling, sensory modalities, etc. (see diagram on page 53), occurring in an accurately describable form. However, the I Ching insists that all 8 spiritual capacities should be related to one another only after very careful thought in pursuance of the injunction: Bring about central harmony = unity with the Tao — the promise of wisdom, harmony, and happiness as falling to one's lot.

9. *Similarity* of the central philosophical importance of the problem: chance and necessity in Monod and the I Ching.

Monod calls for "man to awake so as to become the objective observer of this Nature" in the full knowledge of his fortuitous and now necessarily invariable existence "free from animistic dreams".

I Ching is the manifestation of early science (= objectivity) with a world theory, acceptable to every modern physicist, of plus and minus charges, quanta, etc. terminating in congruence with the DNA as described in the foregoing.

We cannot imagine this knowledge being acquired without "objective" and therefore scientific observation of Nature in early history, although, of course, we cannot tell how such observations were made.[2]

10. *Similarities* as phenomenon. In the I Ching, the "standard work" of Chinese science, as a modern Chinese said in reply to the question why such an intelligent people had not produced any natural science comparable to that of the West[3], there is a curious unity of book, theme, reality, and of nature, spirit and intellect which goes beyond a mere book. A book that functions as "actus", which not only points to something but *is* at the same time this "something".

We have exactly the same simultaneous representation of code and coded form, of structural plan and building, in the DNA. (In each cell there is the whole structural plan, the building consisting of trillions of structural plans).

We are no doubt justified in recording this existential and phenomenal *similarity* in support of our thesis.

11. Meditative consummation, possibly identification of the two "codes". In the light of these *similarities* the theme is presented to the reader in a form of meditation which is not doubt unusual but nevertheless appropriate to the Far Eastern subject, being based on the freely varied 5-stage system of Buddhism (it was with this method that Master Dung Shan led his pupils to knowledge).[4] The reader is invited to practise this on his own, whether or not with success he will see for himself. He may ponder similarity, parallels, the hypothesis that the two codes are identical, perhaps some stimulating idea that may occur to him, or even the results of freely ranging imagination.

12. The hypothesis reads: The I Ching, which is accredited by its "emergence" in the DNA and the statistical significance of a large number of *similarities,* contains a world formula with the stature of an order of reality. Is the answer to Heisenberg's quest for those "anonymous basic forms and polar symmetries of uniform Nature"[5] and Monod's puzzlement over the "ultimately unresolved riddle of the origin of the genetic code" 5000 years old?

For the commentary to the last predictive symbol No. 64 of the I Ching closes with the words: "The Book of Changes is a Book of the Future!"

Epilogue
I Ching and DNA
An Interdisciplinary Phenomenon
by Frank Fiedeler Ph.D.

In America it was the molecular biologist Gunther S. Stent who first noticed the congruence between I Ching and DNA and published it in 1969 in his book "The Coming of the Golden Age", He writes:

"The I Ching is based on the interaction of the two antithetical principles Yang (represented by an unbroken line —) and Yin (represented by a broken line - -). Yang and Yin are combined to form four diagrams, Old Yang (=), Old Yin (= =), New Yang (==) and New Yin (==), and the four diagrams are combined three at a time to form $4^3 = 64$ hexagrams. Each hexagram, which is read from bottom to top, represents one of 64 fundamental aspects of life, the nature of each aspect being given by the interaction of the three diagrams of which the hexagram is composed. During the long history of the I Ching, the hexagrams have been arranged in several different ways, of which a so-called "natural" order was worked out during the Sung period about a thousand years ago..."

After some remarks at this point on Leibniz and the relationship between the I Ching and the binary number system, Stent proceeds:

"But however surprising may be the anticipation of binary digits by the I Ching, the congruence between it and the genetic code is nothing short of amazing. For if Yang (the male, or light, principle) is identified with the purine bases and Yin (the

female, or dark, principle) with pyrimidine bases, so that Old Yang and Yin correspond to the complementary adenine (A) and thymine (T) pair and New Yang and Yin to the complementary guanine (G) and cytosine (C) pair, each of the 64 hexagrams comes to represent one of the nucleotide triplet codons. The "natural" order of the I Ching can now be seen to generate an array of nucleotide triplets in which many of the generic codon relations manifest in Crick's arrangement are shown. Perhaps students of the presently still mysterious origins of the genetic code might consult the extensive commentaries on the I Ching to obtain some clues to the solution of their problem."[1]

The coordination of the four DNA bases with the four I Ching diagrams which Stent proposes here is at variance with the coordination which Schonberger — who was unaware of Stent's work — gives in this book.

Schonberger also coordinates adenine with ▬▬ and thymine with ▬ ▬ but makes cytosine correspond with ▬ ▬ and guanine with ▬▬. It is perhaps impossible at the moment to say with abolute certainty which of these coordinations is correct. There has so far been no comparative study in detail of the 64 meaning or function complexes arising on both sides. Moreover, there are in theory six other coordinations which are possible withouth infringing the principle of complementarity in pairs to be found both in the bases and the diagrams. So far as I can see, however, everything at present seems to argue in favor of Schonberger's coordination — although Stent seems at least to have logic on his side in that he coordinates the two purine bases with the two yang symbols and the pyrimidine bases with the two yin symbols. For appearances are deceptive here, the reason being a very commonplace one. Professor Stent no doubt obtained his knowledge of the I Ching diagrams from the American translation of the German translation of the original Chinese text by Richard Wilhelm. Now, presumably because of a printing error, precisely the two diagrams "young yang" and "young yin" have been confused[2]. Actually "young yang" (shao-yang 少 陽) is the Chinese name of the diagram ⚏, "young yin" (shao-yin 少 陰) the name of the diagram ⚎ [3], in other words, the reserve of Stent's nomenclature! It follows from this that it is in fact not Stent's but Schonberger's coordination which corresponds to Stent's intended logic. Further, a first look at the meaning complexes deriving therefrom seems to show that Shconberger is right. Thus, for example, the transcription of the triplet TGA, which as a "punctuation codon" represents a chemical stop signal, is in Schonberger's coordination the hexagram No. 12 ䷋ Pi 否 "Standstill" whereas in Stent's coordination it is the nexagram No. 53 ䷴ Chien 漸 "Development."

A closer examination of the semantic coincidence between the individual words of the two code systems would call for a detailed interdisciplinary investigation. It is questionable whether our present knowledge would suffice at all for such an undertaking. I am not in a position to answer this question as an expert where DNA research is concerned. However, as to the sinological establishment of the original form of the I Ching casuistry, it must be said that — gauged by both the number and also the systematic level of the Chinese texts — this is still today in its very first stages.

Apart from this, however, there are, if only on a purely formal level, such far-eaching parallels in regard to the overall structure that no possible doubt can remain as to the significance of the congruence phenomenon. Let us take, say, the "natural arrangement" of the hexagram to which Stent refers and which the philosopher Shao Yung 邵 雍 (1011-1077) developed on the model of a classical formula which showed a lesser degree of differentiation, namely the triagram formula Hsien-t'ien pa kua

先天八卦. This cyclic formula is so constructed that the *complementary* hexagrams are located exactly opposite each other on the circle. What it illustrates, then, is nothing other than the principle of complementarity which also determines the coupling of the two strands of the DNA double helix.

Several of these cyclic symbol formulae have been handed down. In traditional China their structural theory was held to be the strictest formal level of philosophy. In my book "Die Wende" (The Turn) I have derived these formulae cosmologically fromt he classical commentaries of the I Ching and demonstrated their functional correlation in terms of an anthropological approach to the subject based on linguistic theory[4]. Their contexture in its totality yields nothing other than the structural pattern of the basic genetic process of *self-reproduction*, in the encoding and control of which the very special capacity of the genetic material DNA resides.

The phenomenon of congruence is demonstrated with particular clarity by a comparison of the *numerical relationships*. While I was engaged for many years as a sinologist in the study of the I Ching and the theory of its system, there was one important structural factor that was always obscure to me, namely the so-called Chia-tzu 甲子-cycle, a sixty-element series of symbol combinations which in Chinese oracular literature is currently coordinated with the system of the hexagrams. However, a letter from the Marburg molecular physicist and DNA specialist Fritz Popp on the numerical relationships of DNA afforded me a direct lead to the meaning of this series of symbols: The numerical relationship which it represents as a current combination of a ten-element and a twelve-element series of symbols corresponds *in the double helix of the DNA precisely to the common period of counting by turns and counting by triplets*. Two triplets, i.e. six base pairs, contain half a turn, which embraces five base pairs. Since in the I Ching the unit corresponding to a base pair is represented by *two* graphic elements and consequently the three-element triplets correspond to six-element hexagrams, this ratio of 5:6 is also reflected in the double form 10:12 of the Chia-tzu cycle. The series of the sixty symbol combinations of this cycle thus means in the structure of the DNA three whole turns or ten triplets, in other words, the period where counting by triplets and counting by turns coincide at the same point.

In the cosmology of the I Ching the total structure of the 64 hexagrams is traced back to the cosmic information pattern of the so-called Metonic cycle (Chinese: chang 章), a period of very precisely 19 solar years which was known even to the Babylonians and served in China as a base for the calendar from the earliest times[5]. The Metonic cycle is, expressed briefly, the period during which the sun and moon go through the range of their possible combinations before returning to the point of departure. If we compare this with the numerics of the (extremely light-sensitive) DNA, we find that this cosmological coordination is actually verified in that *one period of 19 turns in the double helix when counted by triplets coincides precisely with the sixty-fourth triplet and thus marks the totality of the 64-element code.* Nineteen complete turns are 190 base pairs, 64 triplets 192 base pairs. The missing two base pairs can be explained by reference to a kind of overlapping principle which figures prominently in the theory of symbol formation in the I Ching[6]. According to this the first two and the last two base pairs of each sequence of 64 triplets would also count as the last and first members of the preceding and following sequence, as a result of which each sequence seems to be two base pairs short in the total count.

With regard to the constitution of the DNA it seems reasonable to conclude that *one turn of the double helix originally represented the information unit of one year.* The genetic material contained in each cell of the human body consists of billions of such turns. Thus the turns of the DNA are evidently like annual rings and represent the annals of the race which have been continued since the dawn of history.

The phenomenon of the structural analogy between I Ching and DNA will have important consequences for the basic theory of the anthropological sciences. It points the way to a *genetic anthropology* which, once followed, will supply the groundword for a science of culture which a truly objective foundation in the natural sciences. Clearly, it is only in this century that physics, through its investigation of DNA, has reached a level at which it can provide such a foundation. Thus a possibility has been realized which has been repeatedly and dogmatically denied by the metaphysics of the moral sciences which, inspired by theology, has been fixated on the absolute special status of human consciousness.

The Opinion of an Elementary Particle Physicist - F. A. Popp

From the discovery of the Genetic Code by Watson and Crick up to genetic engineering, the current development cycle has traveled a straight, although difficult road, trodden by legions of specially trained scientists, paved with untold millions of research funds, admired by an awed public, promising, but also frightening because no one really knows where this road will lead.

Modern molecular biologists work on the patchwork of our material existence with self-focusing precision. Insofar as life is proven to be a space and time-spanning phenomenon based on the dance of the molecules, they break through the borders between biological material and personality, and even life itself, with the confidence of a programed monster.

In this situation a single physician posed a critical question: Is there a connection between this substantiated reflection of biological evolution and the purpose, the indication and the meaning of life?

This question doesn't destroy the barrier between physical material and the spirit, but rather seeks to conquer the barrier using the instrument of the spirit itself. It focuses our curiosity on the phase-border between the syntax and semantics of the DNA structure.

The author presents a proposed solution in an independent manner which might seem even humorous to some due to its unfamiliar form, to another as a genial solution, to still a third considered as an arrogant error. In the analogy of the I Ching to the syntax of the genetic code, Martin Schönberger recognizes the encyclopedia of biological evolution, the "hidden key to life."

This insight is not to be over estimated because through unsuccessful attempts to objectify science, dangerous powers with terrible consequences are being (and will be in the future) built. Consideration of the limitations of and appropriateness to the reference system "humanity" lead to the necessary humility and respect.

On the other hand, the paradigm introduced here leads to the recognition that spirit can be transformed from matter, and matter only gains in significance through the spirit. An apparently elementary, complementary truth stands behind this transformation. The genetic I Ching provides such a "periodic system of the spiritual element," and acts as a goal for human evolution.

BIBLIOGRAPHY

Fiedeler, Frank: *Die Wende*. Berlin: Kristkeitz 1977.

Friedlander, S. H.: *Schopferische Indifferenz*. Munchen: E. Reinhardt 1926.

Gebser, Jean: *Der unsichtbare Ursprung*. Freiburg: Walter 1971.

Gebser, Jean: *Dualismus und Polaritat*. Zurich: Pestalozzi 1971.

Granet, Marcel: *Das chinesische Denkin — Inhalt, Form, Charakter*. Munchen: R. Piper 1971.

Gundert, Wilhelm: *Bi-Yan-Lu* (2 Bde.). Munchen: C. Hanser 1960/67.

Heisenberg, Werner: Die Einheit der Natur bei A. v. Humbolt und in der Gegenwart, *Suddeutsche Zeitung*, Munchen, Nr. 310 v. 27./28. 12. 1969.

Hummell, Siegbert: *Polaritat in der chinesischen Philosophie*. Leipzig: O. Harrassowitz 1949.

Hummel, Siegbert: *Zum ontologischen Problem des Dauismus (Taoismus)*. Leipzig: O. Harrassowitz 1948.

Kellerer, Christian: *Objet trouve, Surrealismus, Zen*. Reinbek: Rowohlt 1968.

Lassalle, H. M. Enomiya: *Zen — Weg zur Erleuchtung*. Wien: Herder 1969.

Leibniz, G. Wilhelm: *Zwei Briefe uber das Binare Zahlensystem und die chinesische Philosophie*. Munchen: Belser 1968 (Herausgeber und Nachwort: Jean Gebser: Zur 5000jahrigen Geschichte des Binaren Zahlensystmes: Fu-Hi — G. W. Leibniz — Norbert Wiener).

Monod, Jacques: *Chance and Necessity*. Random House 1972.

Watts, Allan W.: *Natur — Mann und Frau*. Koln: DuMont Schauberg 1972.

Wilhelm, Richard/Baynes, Cary *The I Ching* Princeton, Princeton University Press 1967.

Wilhelm, Richard/Jung, C.G. *The Secret of the Golden Flower* New York, Harcourt, Brace & World 1962.

Yuan-Kuang: *I-Ging*. Planegg: O. W. Barth 1951.

Footnotes

Stone Rubbing

[1] Burckhardt, Erwin: "Chinesische Steinabreibungen", Hirmer Verlag, Munich 1961, plate 1.
[2] Granet, Marcel: "Das Chinesische Denken — Inhalt, Form, Charakter", R. Piper Verlag, Munich 1971, p. 134 ff. and 273 ff.
[3] Chavanne's, Edouard: "Les Memoires Historiques de Se-Ma-Ts'ien", Ernest Leraux, Paris, 1895, p. 5ff.

Foreward to the Second Edition

[1] Dr. H. de Witt: Analogik. Vol. I, 315 pages and 25 pages index. Basel: Wepf & Co. Verlag. Vol. II in preparation.

[2] Gebser, Jean: Ursprung and Gegenwart. Munich; dtv 1973.

[3] Chargaff, Erwin: Nutzliche Wunder. Gedanken uber die Nubleinsaureforschung, in Scheidewege, Vol. 6, Fasc. 3, 1976. Stuttgart: Klett Verlag.

Foreword

[1] The Essay appeared in Bitter, W.: "Dialog uber den Menschen", Klett Verlag, Stuttgart 1968 (pp. 231 ff and 249 ff).

[2] Monod, Jacques: "Chance and Necessity"

[3] op cit.

The Discovery

[1] Reprint of the first publication of the "Discovery" in the Zeitschrift fur Allgemeinmedizin — Der Landarzt, No. 16/69, as a reply to an article by Grafe in No. 5/69 of the same journal.

The Code of the I Ching

[1] Wilhelm, R./Jung, C. G.: "The Secret of the Golden Flower".
[2] Wilhelm, R.: "I-Ging, Das Buch der Wandlungen", Eugen Diedrichs Verlag, Dusseldorf and Cologne 1923. English translation ("I Ching or book of changes") by Cary F. Baynes, Princeton University Press, Princeton, N.J.
[3] Wilhelm, R.: "The Book of Changes", Vol. 1, VIII: "This fundamental postulate is the great primal beginning of all that exists, *T'ai chi* — in its original meaning, the "ridgepole".
[4] Hummel, Siegbert: "Polaritat in der Chinesischen Philosophie", Otto Harrassowitz Verlag, Leipzig 1949.

Polarity in The I Ching and The Genetic Code

[1] As the sinologist Marcel Granet reports ("Das Chinesische Denken-Inhalt, Form, Charakter", R. Piper Verlag, Munich 1963, p. 135) Fu-Hsi is depicted "in a mutual embrace with his wife, their bodies tapering off into entwined snakes (!)".
[2] Friedlaender, S.H.: "Schopferische Indifferenz", Reinhardt Verlag, Munich 1926.
[3] Gebser Jean, Ursprung und Gegenwart, DTV. Verlag, Munich 1973.
[4] Watts, Alan W.: "Nature — Man and Woman"

I Ching The World Code and DNA The Life Code — A Key?

[1] A formal analysis by the author; not sinologically substantiated.
[2] Yuan-Kuan: "I-Ging", O. W. Barth Verlag, Munchen-Planegg 1951

I Ching, Law of Chance

[1] Wilhelm, R.: I-Ging. Das Buch der Wandlungen, Diederichs Verlag, Dusseldorf und Koln 1923, Vol. I, pp. 280 ff.
[2] Wilhelm R. / Jung C.G.: "The Secret of the Golden Flower," p. XIII.

Chance and Necessity in DNA, Surrealism and I Ching

[1]Monod Jacques: "Chance and Necessity"
[2]Quoted from a report on "Chance in Evolution", Suddeutsche Zeitung, 3/4 June 1972, report on the meeting of Nobel laureates.
[3]Monod, J.: "Chance and Necessity", p. 82, Maxwell's demon
[4]Kellerer, Christian: Objet trouve, Surrealismus, Zen", Rowohlt Taschenbuch Verlag, Reinbek b. Hamburg 1968
[5]For precise formulations see p. 48 ff.

The Practice of Prediction in the I Ching

[1]Wilhelm, R./Jung, C. G.: "The Secret of the Golden Flower" p. XII.
[2]Paracelsus and Heisenberg are aware of this. It is to scientists like Heisenberg that the credit goes for overcoming the subject-object dualism in science.

The I Ching and The Genetic Code in the 5-Stage Pattern of Meditation

[1]Bi-Yan-Lu, Meister Yuan-wu's Niederschrift von der Smaragdenen Felswand ... translated into German and explained by Gundert, Wilhelm, Carl Hanser Verlag, Munich 1967, Vol. II, p. 213 ff.
[2]Dumoulin, H.: "Zen-Geschichte und Gestalt", Francke Verlag, Berne 1959, p. 118 ff.
[3]Prof. Margaret O. Dayhoff, Silver Springs/Md.
[4](Dept. for Electronic Data Processing, University of Giessen) Graul/Francke: "Kunftige Entdeckungen der Physik", Zeitschrift fur Allgemeinmedizin, 32/1970, p. 1591.
[5]Suddeutsche Zeitung, Munchen, No. 249, 17/18 October 1970.
[6]See Bibliography
[7]23.9.38 Schloss Nymphenburg, Munich.
[8]Friedlander, S.H.: "Schopferische Indifferenz", Ernst Reinhardt Verlag, Munich 1926
[9]Groddeck, Georg: "Das Buch vom Es", Limes Verlag, Wiesbaden 1961.
[10]Gebser, Jean: "Dualismus und Polaritat", in "Ursprung und Gegenwart," DTV Verlag.

Summary of All Reflections

¹See Watson's report (Watson, James D. *The Double Helix,* Penguin Books) on the discovery of the double helix—he had a genius-like imagination which enabled him to win the race!

²But the I Ching also demands that this observer be made into the whole and complete subject of all objects by the 7 other functions of feeling and intuition, all of which are to be exercised with the same objectivity (this is no doubt the objective path of the wise man of the I Ching!). Unless all the 8 functions of the mind in their polar pattern are uniformly cultivated (see p. 50) there can be no true objectivity. An example: construction of the atomic bomb without consideration for feeling, sensation, intuition, polar humanity, religion — —, here there was only ice-cold intellect, the dualistic distortion of reason at work (without due attention to the other pole = intuitive grasp of the whole human situation) coupled in the event with an equally ice-cold determination to carry the project through (dualistic distortion and neglect of the other pole — the warmly developed life of feeling). Only if the pairs are fully respected can they be truly human!

The reader is asked to bear with this fully worked example of complete objectivity as a reply in terms of the I Ching to Monod's highly partial "objectivity".

³C.G. Jung, "The Secret of the Golden Flower", p. XI.
⁴See page 109
⁵See page 107

Epilogue

¹Gunther S. Stent, The Coming of the Golden Age, New York 1969, p. 64/65

²Richard Wilhel, The I Ching or Book of Changes, pub. London Routledge & Kegan Paul

³Vide e.g. Fung Y-lan, A History of Chinese Philosophy, Princeton 1953, Vol. II, p. 459

⁴Frank Fiedelger, Die Wende, Ansatz einer genetischen Anthropologie nach dem System des I-ching, Berlin 1974

⁵Cf. Frank Fiedeler, Die Wende, Chapter V

⁶Cf. Frank Fiedeler, Die Wende, Chapter III a.

Other Titles From Aurora

THE PULSE IN OCCIDENT AND ORIENT
Reuben Amber

AWAKEN HEALING ENERGY THROUGH THE TAO
Mantak Chia

ASTROLOGY OF PERSONALITY
Dane Rudhyar

SELF HEALING, YOGA & DESTINY
Elisabeth Haich

HOW COSMIC & ATMOSPHERIC ENERGIES INFLUENCE YOUR HEALTH
Dr. Michel Gauquelin

For a complete catalog write:

**AURORA PRESS
P. O. BOX 573
SANTA FE, NM 87504**